SpringerBriefs in Environmental Science present concise summaries of cutting-edge research and practical applications across a wide spectrum of environmental fields, with fast turnaround time to publication. Featuring compact volumes of 50 to 125 pages, the series covers a range of content from professional to academic. Monographs of new material are considered for the SpringerBriefs in Environmental Science series.

Typical topics might include: a timely report of state-of-the-art analytical techniques, a bridge between new research results, as published in journal articles and a contextual literature review, a snapshot of a hot or emerging topic, an in-depth case study or technical example, a presentation of core concepts that students must understand in order to make independent contributions, best practices or protocols to be followed, a series of short case studies/debates highlighting a specific angle.

SpringerBriefs in Environmental Science allow authors to present their ideas and readers to absorb them with minimal time investment. Both solicited and unsolicited manuscripts are considered for publication.

More information about this series at http://www.springer.com/series/8868

Koyel Sam • Namita Chakma

Climate Change in the Forest of Bengal Duars

Response of Life and Livelihoods

 Springer

Koyel Sam
Department of Geography
Dr. B.N.D.S. Mahavidyalaya
Purba Bardhaman, West Bengal, India

Namita Chakma
Department of Geography
The University of Burdwan
Burdwan, West Bengal, India

ISSN 2191-5547   ISSN 2191-5555   (electronic)
SpringerBriefs in Environmental Science
ISBN 978-3-030-73865-5   ISBN 978-3-030-73866-2   (eBook)
https://doi.org/10.1007/978-3-030-73866-2

This Springer imprint is published by the registered company Springer Nature Switzerland AG
The registered company address is: Gewerbestrasse 11, 6330 Cham, Switzerland

Dedicated to Professor Sunando Bandyopadhyay

Preface

Presently, climate change is a stark reality associated with vulnerability faced by living beings all around the world and therefore has gained importance in the field of scientific research. The Eastern Himalayan mountain ecosystem is fragile in nature, and the phenomena of climate change have imposed extra vulnerable conditions in this unique landscape. Adjacent to the Eastern Himalaya, the foothill landscape is also hazard-prone. The present study area is the Bengal Duars region—a foothill landscape of the Eastern Himalaya. The book will focus on the climate change and 'struggle for existence' issues facing the forest villagers to the changing climatic situation.

This book encompasses seven chapters. Each chapter starts with an abstract describing, in brief, the theme of the chapter. Chapter 1 provides an overview about the study, including a review of the literature, significance, objectives, and methodology of the research. Chapter 2 illustrates the study of the area with its unique physio-social characteristics. Chapter 3 describes the forestry of Bengal Duars. Chapter 4 estimates present and future climate change in the context of the Bengal Duars region. Both temperature and rainfall are taken under consideration to identify more than 100 years' climatic trends and oscillations. Chapter 5 evaluates the spatial vulnerability of forest from the villages in Bengal Duars in continuation with the climate change under the domain of sensitivity, exposure, and adaptive capacity. Chapter 6 assesses ground observation of the forest villager's perception and responses about changes and impacts that they have faced. Analytical techniques are applied for critical evaluation and assessment of livelihood crises. Chapter 7 evaluates adaptive attitudes of the forest villagers in order to judge their resilience capability. In this regard, a conceptual model has been developed by interlinking their concern, household profile, livelihood assets with potential, and actual adaptation attitudes. Finally, Chapter 8 illustrates recommendations and suggestions for the restoration of this unique landscape.

West Bengal, India

Koyel Sam
Namita Chakma

Acknowledgements

We are indebted to acknowledge the India Meteorological Department (IMD), Pune, for providing climate gridded datasets which are used in the present study for a long-term climate change analysis.

We are grateful to the officials of the West Bengal Forest Department and Divisional Forest Range of Alipurduar and Jalpaiguri Divisions for providing relevant information for the study. We are also thankful to the people of the forest villages for their help during the field survey. We would like to thank M.A./M.Sc. students of the Major Elective Course 'Regional Planning and Development', Department of Geography, the University of Burdwan (session 2016–2018) for their help during the field survey (2017–2018).

We are also thankful to The University of Burdwan, Purba Bardhaman, West Bengal, for ensuring various facilities to carry out the research by providing access to library facilities, instruments, software, maps, etc., which are used in various stages of the study.

We are thankful to Herbert Moses and Zachary Romano (Project coordinator, Book, Springer) for their constant guidance to complete the work in time.

Finally, we would like to express our heartfelt gratitude to our family members for their generosity and support.

Koyel Sam
Namita Chakma

Contents

Abbreviations

AHP	Analytical Hierarchy Process
APL	Above poverty level
BPL	Below poverty level
BTR	Buxa Tiger Reserve
CMIP5	Coupled Model Inter-comparison Project Phase 5
CWC	Central Water Commission
FAO	Food and Agriculture Organization
FPC	Forest Protection Committees
FRL	Forest Reference Level
FSI	Forest Survey of India
GCMs	General Circulation Models
IMD	India Meteorological Department
IPCC	Intergovernmental Panel on Climate Change
IUCN	International Union for Conservation of Nature
JFM	Joint Forest Management
MGNREGA	Mahatma Gandhi National Rural Employment Guarantee Act
MODIS	Moderate Resolution Imaging Spectroradiometer
MoEFCC	Ministry of Environment, Forest and Climate Change
NAPCC	National Action Plan on Climate Change
NCEF	National Clean Energy Fund
NFMS	National Forest Monitoring System
NS/AC	National Strategies/Action plan
PCA	Principle component analysis
RCMs	Regional climate models
REDD	Reducing emission from deforestation and forest degradation
ROAM	Restoration Opportunities Assessment Methodology
SIS	Safeguard Information System
SOI	Survey of India
UNDP	United Nations Development Programme
UNEP	United Nations Environment Programme

UNFCCC	United Nations Framework Convention on Climate Change
WMO	World Meteorological Organization
WRI	World Resources Institute

List of Figures

List of Tables

Chapter 1
Introduction

Abstract The Asia-Pacific region is the least forested region (0.2 hectares per capita) in the world, and most of the forest cover is threatened by human disturbances. Bengal Duars, a region well known for its forests and biodiversity lays in the foothills of the Eastern Himalaya. Other than anthropogenic disturbance in forests, climate change and associated vulnerability are becoming a serious threat in such a landscape. According to the climate risk index, most of the Asian countries are in vulnerable condition and will be severe in future due to climate change. The purpose of this chapter is to provide a brief note on the theme of the study, review of literatures on the aspects of climate change, vulnerability and adaptation in the context of Asia. This analysis further helps to identify the research gaps, objectives, and also to frame the methodology accordingly to enhance the relevance of the present study.

Keywords Bengal Duars · Eastern Himalaya · Climate change · Vulnerability · Adaptation

1.1 Introductory Notes

Climate change is now becoming a harsh reality, and its impact is faced by living beings all around the world. The earth's climate is warming rapidly, and scientists are working to find the scenario of changing climate and plan strategies for adapting to climate change in the future. According to the IPCC (2014c) report, the recent warming of climate seems to be unambiguous and most likely because of emission of greenhouse gases by anthropogenic activities. All around the world, forest cover is the most transformed and disturbed landscape due to unscientific activities of humans.

A forest is a natural purifier of air as it has the ability of carbon sequestration, controller of water cycle, and protector of watershed and habitat. Altogether, the forest plays a crucial role in safeguarding environment and biodiversity. Between 1990 and 2015, the global forest area declined to 3.16% due to indiscriminate felling of trees. Asia-Pacific region is the least forested region of the world where per capita forest is 0.2 hectares, whereas South Asia stands at only 0.05 hectares (FAO 2011). As South Asia is the least forested sub-region (0.05 hectare per capita) of the Asia-Pacific region, the pressure on forest resource for competing uses is also high that causes degradation of forested landscape. Forest plays a vibrant role in climate change adaptation efforts by providing food safety, reducing risk of disaster and regulating microclimate. All around the world, about 1.6 billion people depend on forest resource for their livelihood (World Bank 2001). It is evident that climate change is threatening the livelihood of people who are more dependent on natural resources. Therefore, it is challenging to support poor people to reduce poverty especially in rural areas (FAO 2009). In the discourse of human–environment relationship, the understanding of the environment and responses of humans towards environmental change play a significant role.

1.2   Review of Literature

Climate system may be global in extent, but it is manifested by regional and local processes in terms of occurrence, characteristics and implications. Thus, the importance of the regional aspect of climate change has been recognized by the IPCC in its fifth Assessment Report (Part-B). In this section, the literature is reviewed in the context of Asian continent. Aspects like climate change, vulnerability, adaptation, life and livelihoods of people in forests of different countries of Asia have been reviewed.

1.2.1   Climate Change

Globally, the averaged combined warming of land and ocean surface temperature follow the trend of 0.85 °C (0.65–1.06 °C) during the period 1880–2012. Asian countries have experienced warming trends and higher extremes almost all over the continent. In South-East Asia, temperature has been increasing at the rate of 0.14 °C to 0.20 °C/decade and the trend of increasing number of hot days and warmer nights have increased since the 1960s. In North Asia, an increasing trend of heavy precipitation has been observed, but inter-decadal variability, more frequent deficit monsoons under inhomogeneity has been found in the South Asia (IPCC 2014b).

In the West Asian Country of Oman, yearly rainfall is quite variable and irregular, and experiences positive relationship with topography (Kwarteng et al. 2009). Myanmar exhibited a 2–6-year cycle of variability of rainfall and an identical change was noted from 1989 onwards (Sein and Zhi 2016). In 1960–2009, South-West China had experienced significant positive trend of maximum and minimum temperature, and decreasing trend for rainfall amount and rainy days (Qin et al. 2015). The long-term warming trend of temperature in Tibetan Plateau was investigated by Frauenfeld et al. (2005) based on station data in 2005. The rise of temperature (0.6–1.0 °C) and decreasing trend of summer rainfall in arid, coastal, mountain and plain areas of Pakistan over 50 years was studied by Farooqi et al. (2005). Trends of temperature and precipitation extremes in Nepal have been investigated by Baidya et al. (2008). From 1971 to 2006, temperature extremes have become higher in mountainous regions compared to Terai belt, and trend of precipitation has increased all around. However, in Bangladesh, monsoon rainfall experienced above normal after 1963 onwards and a similar rainfall variation has been found in Northeast India (Kripalani et al. 1996). The annual average and pre-monsoon rainfall trend has also been found to increase (Shahid 2010).

In India, the amount of overall seasonal mean rainfall has decreased, but an increase in extreme rainfall events has been observed in different parts (IPCC 2014b), and WMO (2019) also reported the occurrence of extreme events in northern and eastern coasts of the Bay of Bengal and the eastern part of Himalaya. Subsequently, sub-Himalayan Bengal has reported an increasing trend of heavy rainfall by Raha et al. (2014), Basak (2014), Varikoden et al. (2013). On the other hand, the warming trend of temperature in India has been revealed through several studies with varying degrees of changes based on long-term analysis (Pant and Kumar 1997; Ravindranath et al. 2011). The multi-model and multiple scenarios of temperature and rainfall of India is likely to project 1.7–2 °C by 2030s and 3.3–4.8 °C by 2080s relative to the pre-industrial times, and precipitation is projected to increase from 4% to 5% by 2030s (Chaturvedi et al. 2012). Based on 30 years' data (1979–2008) of four stations in northern parts of West Bengal, trends of surface temperature and rainfall were investigated by Raha et al. (2014). Significant warming of mean minimum temperature in all the seasons became more pronounced in Jalpaiguri station with an increase of rainy days and high rainfall.

1.2.2   Vulnerability

Vulnerability as a multidimensional facet incorporates various factors responsible for susceptibility to losses. UNEP (2002) defined vulnerability as 'the interface between exposure to the physical threats to human welfare and the capacity of residential areas to cope with those threats'. In the wide-ranging expanse of vulnerability, climate change is a newly emerging threat added in the paradigm of vulnerability research. Regarding the dimension of vulnerability, it can be studied in various ways based on concepts and methods. The conceptual dimension of vulnerability is also useful for framing the issues (Füssel 2007; Ionescu et al. 2009; Soares et al. 2012). Other studies focus on methods and approaches of vulnerability. Econometric and variable assessment approaches help to identify the prioritized risks and vulnerable groups based on socio-economic data to receive policy attention (Pritchett et al. 2000; Prescott-Allen 2001; Heitzmann et al. 2002). Indicator-based approaches exercise several indicators under specific domains and categorized them depending on weightage (Easter 1999; Deressa et al. 2009; Gbetibouo et al. 2010; Piya et al. 2012). The IPCC (AR5) analysed the model of vulnerability as a function of exposure, sensitivity and adaptive capacity (IPCC 2014a).

Yusuf and Francisco (2010) carried out a detailed and intensive research work on *Hotspots! Mapping Climate Change Vulnerability in South-East Asia* in 2010. They were investigating spatial (region/district/ provinces) vulnerability followed by the IPCC framework. High level of vulnerability is found in Indonesia's urban hotspots due to high population pressure. Philippines and Vietnam are highly exposed to climate change and are with low adaptive capacity.

The scarcity of drinking water due to climate change in coastal South and South-East Asia was investigated by Hoque et al. (2016). It assessed spatial vulnerability of drinking water due to salinization caused by meteorological variability and climate change. It has been found that the coastal areas of Vietnam, India and Bangladesh are most vulnerable as 25 million people are at risk to drink 'saline' water.

An attempt was made by Chitale (2014) to study forest climate vulnerability of Chitwan-Annapurna Landscape in Nepal Himalaya. The biophysical variables, MODIS forest observation and socio-economic variables were used to find out spatial vulnerability. They reported that the region located at a high altitude is to be moderate to highly vulnerable in comparison to the Terai region.

The climate change vulnerability of North-East India has been evaluated by Ravindranath et al. (2011). The vulnerability profiles of different sectors (water, agriculture and forest) has been analysed at district level based on present and future climates. Index-based approach and Principal Component Analysis were executed to develop an impact assessment model.

Upgupta et al. (2015) studied vulnerability of forest due to climate change at the state and regional level in the Western Himalaya. Vulnerability of forest was evaluated using indicator-based approach, and future vulnerability was assessed based on climate and vegetation impact models.

The Hadley Centre Global Environment Model (version 2) known as HadGEM2 family of climate models was applied to impart the vulnerability of agro-ecological zones of India in different RCPs scenarios by Shukla et al. (2017). The study revealed that districts located in semi-arid agro-ecological zones of northern plains, western plains and central highlands are the most vulnerable in the current scenario (1950–2000). The study also predicted that districts of Deccan plateau and Central (Malwa) highlands will be highly vulnerable in future (2050).

1.2.3   Adaptation of People

Adaptation is defined by Smit and Pilifosova (2001) as 'response to actual or expected climatic stimuli and their effects or impacts'. Climate change threatens life and livelihoods of people especially in less developed countries with less adaptive capacity (IPCC 2012). Depending upon adaptive capacity of a system, vulnerability may reduce. Adaptive capacity is defined by Adger et al. (2007) as 'ability or potential of a system to respond successfully to climate variability and change'. Adaptive capacity is determined by a range of factors such as social, economic, natural and human capitals as well as values and perception of respondents (Rayner and Malone 1998; Kelly and Adger 2000).

Climate change impacts on agriculture and adaptation in Kazakhstan were studied by Mizina et al. (1999). The Adaptation Decision Matrix used subjective measures to perceive how well adaptation options can meet the policy objectives. It has been observed that though the highest benefit may be gained by controlling soil erosion its implementation appeared as cost-ineffective relatively from free market measure.

Dang et al. (2014) included psychological factors to investigate adaptation intentions of farmers in the Mekong Delta of Vietnam. The study revealed that farmers are more likely to adapt when they perceived high risk of climate change and effectiveness of adaptive measures. Willingness of adaptation increased when they got pressure from friends and neighbours to adapt. They also began to adapt at the time of rising price of electricity, water and fuel.

Application of local knowledge to adapt climate change in Bangladesh was studied by Anik and Khan (2012). It has been revealed that about 10% of the respondents have good knowledge of climate change. Local people are changing their behaviour related to livelihoods with the changing nature of climatic hazards. Out of the 16 adaptive measures, six are found to be the most well-adaptive strategies identified in this study, for example, crop diversity, cage aquaculture, duck rearing, floating gardening, re-digging of canal, construction of embankment and walls.

Gao et al. (2014) studied the impact of climatic variability on the alpine grassland ecosystem and adaptation strategies in Tibetan Plateau. To combat with the future trend of warming, specific adaptation measures like water-saving irrigation, rotational grazing management, grassland fencing and grass planting were proposed in the study.

Banerjee et al. (2016) attempted a study on 'Adaptation strategies to combat climate change effect on rice and mustard in Eastern India'. Crop growth simulation model was calibrated and validated for the said crops for West Bengal to predict yield under thermal conditions of 1–3 °C rise of temperature. It has been found that the negative impact of climate change is more pronounced on mustard crop yield in the winter season than rice. The authors suggested modification in sowing time and increased rate of nutrient application in the field as effective adaptation strategies.

The robust research work on Hindu Kush Himalaya about mountains, environment, climate change, sustainability and people has been conducted by Wester et al. (2019). The region covers seven mountainous least developed countries (excluding Pakistan) of the world. In the case of agriculture, people adopted certain strategies like use of water-saving technology, change of crop calendar, practice of terrace farming, cultivation of less water-intensive crops and irrigated farming. Indigenous communities are applying local ecological knowledge and traditional techniques for terracing and stabilization of slopes with plants. Seasonal and daily migration are also identified as a way of diversifying livelihood for a secure source of income that is not affected by any shocks.

There has been some significant research work about climate change impacts on Asian countries specifically in India. Most of the above-stated studies are very useful for comparative assessment of relative changes in climatic variables, level of vulnerability, and ways of adaptation of people, but most of them have used model-based approaches like General Circulation Models (GCMs) to find the impacts. There are few research studies at micro scale or household level to assess the in-depth impacts of climate change on life and livelihoods. After a thorough review of the available literatures, it has been identified that there exist significant research gaps to explore the impact of climate change with associated vulnerability in forests of the Bengal Duars region.

1.3 Objectives

The objectives of the present study are as follows:

1. To study the spatial extension and evolution of Bengal Duars region
2. To estimate long-term (more than 100 years) changes in climate
3. To evaluate spatial vulnerability in the forested landscape
4. To assess people's perception on changing climate and its impact on life and livelihoods
5. To assess sustenance and adaptive attitudes of forest villagers
6. To examine the role of governmental policy and suggest some measures for the management of forested landscapes in the context of climate change

1.4 Methodology

The study being proceeded according to the scheduled phases as follows (Fig. 1.1):

Phase 1: Necessary information and data have been collected from different sources in order to understand the geographical identity and administrative decisions in the evolution of Bengal Duars region. This region itself is well recognized for the richness of forest resources and biodiversity. The evolutionary history depicts how commercialization of forest resources and its exploitation had started during the British period and is still continuing.

Phase 2: Long-term data (more than 100 years) on climatic variables (temperature and rainfall) has been collected from the India Meteorological Department (IMD), Government of India. The trend of climatic variables and the future possibilities have been estimated accordingly.

Phase 3: The forest is a natural regulator to control micro-climate of a region. Besides this, they provide fodder, fuel wood and other forest products to the forest-dependent people. Climate change and climatic variability is a new inducement to accelerate the level of vulnerability in Bengal Duars where people living in forest villages are already struggling due to degradation of forested landscape. Assessment of vulnerability in terms of sensitivity, exposure and adaptive capacity of forest villagers is, therefore, considered in the third phase of analysis.

Phase 4: The impact of climate change and associated vulnerability on the life and livelihoods of forest villagers are analysed in detail on the basis of their perception and responses in this phase.

Phase 5: Finally, management strategies are proposed for the forested landscape of Bengal Duars with a special reference to REDD+ mechanism.

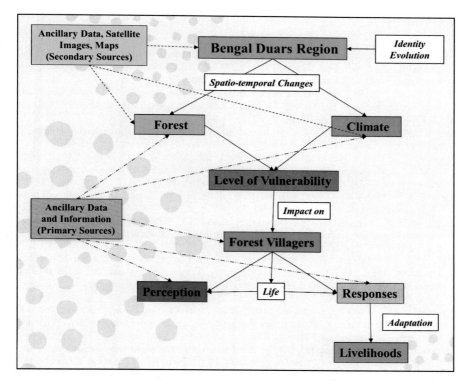

Fig. 1.1 Flow chart of methodology

1.5 Significance of the Study

Climate change is a severe reality faced by the living beings in all parts of the world but with a diverse intensity. Climate is an integral natural factor of a region and change in its variables affect the ecology and environment of a region. Bengal Duars is a natural geographical region known as a biodiversity hotspot of West Bengal, India. Degradation of forest and changing climate raises the issue pertaining to the long-term survival of endemic species and human life.

The present study is relevant in several contexts. *Firstly,* chronological evolution and delineation of this unique geographical region has not been explored systematically before. *Secondly,* measurement of spatial vulnerability of forested landscape may help to understand severity of the issue. *Thirdly,* analysis of human–environment relationship at micro-level with the changing climate will help to understand climate change-related dynamics in other regions also facing similar kind of problems.

References

Adger, W. N., S. Agrawala, and M. Mirza. 2007. Assessment of adaptation practices, options, constraints and capacity. ed. Climate Change 2007: *Climate Change Impacts, Adaptation, and Vulnerability*, chap 17. Cambridge: Cambridge University Press.

Anik, S.I., and M.S.A. Khan. 2012. Climate change adaptation through local knowledge in the north eastern region of Bangladesh. *Mitigation and Adaptation Strategies for Global Change* 17: 879–896. https://doi.org/10.1007/s11027-011-9350-6.

Baidya, S.K., M.L. Shresth, and M.M. Sheikh. 2008. Trends in daily climatic extremes of temperature and precipitation in Nepal. *Journal of Hydrology and Meteorology* 5 (1): 38–51.

Banerjee, S., S. Das, A. Mukherjee, A. Mukherjee, and B. Saikia. 2016. Adaptation strategies to combat climate change effect on rice and mustard in Eastern India. *Mitigation and Adaptation Strategies for Global Change* 21: 249–261. https://doi.org/10.1007/s11027-014-9595-y.

Basak, P. 2014. Variability of southwest monsoon rainfall in West Bengal: An application of principal component analysis. *Mausam* 65 (4): 559–568.

Chaturvedi, R.K., J. Joshi, M. Jayaraman, G. Bala, and N.H. Ravindranath. 2012. Multi-model climate change projections for India under representative concentration pathways. *Current Science* 103 (7): 791–802.

Chitale, V.S. 2014. *Forest Climate Change Vulnerability and Adaptation Assessment in Himalayas*. The International Archives of the Photogrammetry, Remote Sensing and Spatial Information Sciences, Volume XL-8, 2014 ISPRS Technical Commission VIII Symposium, 9–12 December 2014, Hyderabad: India.

Dang, H.L., E. Li, I. Nuberg, and J. Bruwer. 2014. Understanding farmers' adaptation intention to climate change: A structural equation modelling study in the Mekong Delta, Vietnam. *Environmental Science and Policy* 41: 11–22.

Deressa T.T., R. M. Hassan, and C. Ringler. 2009. *Assessing Household Vulnerability To Climate Change The Case Of Farmers In The Nile Basin Of Ethiopia*, IFPRI Discussion Paper 00935, Environment and Production Technology Division, International Food Policy Research Institute.

Easter, C. 1999. Small states development: a commonwealth vulnerability index. *The Round Table* 88 (351): 403–422. https://doi.org/10.1080/003585399107947.

FAO. 2009. *Rural Poverty and Natural Resources: Improving Access and Sustainable Management*. Background Paper for IFAD 2009 Rural Poverty Report. ESA Working Paper No. 09–03. Food and Agriculture Organization of the United Nation.

———. 2011. *Southeast Asian forests and forestry to 2020: Sub regional report of the second Asia-Pacific forestry sector outlook study*. Bangkok: Food and Agriculture Organization of the United Nation.

Farooqi, A.B., A.H. Khan, and H. Mir. 2005. Climate change perspective in Pakistan. *Pakistan Journal of Meteorology* 2 (3): 11–21.

Frauenfeld, O.W., T. Zhang, and M.C. Serreze. 2005. Climate change and variability using European Centre for Medium-Range Weather Forecasts reanalysis (ERA-40) temperatures on the Tibetan Plateau. *Journal of Geophysical Research* 110. https://doi.org/10.1029/2004JD005230.

Füssel, H.M. 2007. Vulnerability: a generally applicable conceptual framework for climate change research. *Global Environmental Change* 17 (2): 155–167.

Gao, Q., Y. Li, H. Xu, Y. Wan, and W. Jiangcun. 2014. Adaptation strategies of climate variability impacts on alpine grassland ecosystems in Tibetan Plateau. *Mitigation and Adaptation Strategies for Global Change* 19: 199–209. https://doi.org/10.1007/s11027-012-9434-y.

Gbetibouo, G.A., C. Ringler, and R. Hassan. 2010. Vulnerability of the South African farming sector to climate change and variability: an indicator approach. *Natural Resource Forum* 34: 175–187. https://doi.org/10.1111/j.1477-8947.2010.01302.x.

Heitzmann, K., R. S. Canagarajah, and P. B. Siegel. 2002. *Guidelines for assessing the sources of risk and vulnerability*. Social Protection Discussion Paper Series 0218. World Bank, Washington, DC.

Hoque, M.A., P.F.D. Scheelbeek, P. Vineis, A.E. Khan, K.M. Ahmed, and A.P. Butler. 2016. Drinking water vulnerability to climate change and alternatives for adaptation in coastal South and South East Asia. *Climatic Change* 136: 247–263. https://doi.org/10.1007/s10584-016-1617.

Ionescu, C., R.J.T. Klein, J. Hinkel, K.S. Kavi Kumar, and R. Klein. 2009. Towards a Formal Framework of Vulnerability to Climate Change. *Environmental Modelling and Assessment* 14: 1–16. https://doi.org/10.1007/s10666-008-9179-x.

IPCC. 2012. *Managing the risks of extreme events and disasters to advance climate change adaptation.* ed. C. B. Field, V. Barros., T. F. Stocker., D. Qin., D. J. Dokken., K. L. Ebi., M. D. Mastrandrea., K. J. Mach., G-K. Plattner., S. K. Allen., M. Tignor., P. M. Midgley. Special report (of working groups I and II) of the intergovernmental panel on climate change. Cambridge and New York: Cambridge University Press.

———. 2014a. *Climate Change 2014: Impacts, Adaptation, and Vulnerability. Part A: Global and Sectoral Aspects. Contribution of Working Group II to the Fifth Assessment Report of the Intergovernmental Panel on Climate Change* ed. V.R. Barros, C.B. Field, D.J. Dokken, M.D. Mastrandrea, K.J. Mach, T.E. Bilir, M. Chatterjee, K.L. Ebi, Y.O. Estrada, R.C. Genova, B. Girma, E.S. Kissel, A.N. Levy, S. MacCracken, P.R. Mastrandrea, and L.L. White. Cambridge: Cambridge University Press.

———. 2014b. *Climate Change 2014: Impacts, Adaptation, and Vulnerability. Part B: Regional Aspects. Contribution of Working Group II to the Fifth Assessment Report of the Intergovernmental Panel on Climate Change,* ed. V.R. Barros, C.B. Field, D.J. Dokken, M.D. Mastrandrea, K.J. Mach, T.E. Bilir, M. Chatterjee, K.L. Ebi, Y.O. Estrada, R.C. Genova, B. Girma, E.S. Kissel, A.N. Levy, S. MacCracken, P.R. Mastrandrea, and L.L. White. Cambridge: Cambridge University Press.

———. 2014c. Synthesis report: *summary for policymakers.* In *Climate Change 2014: impacts, adaptation, and vulnerability. Part A: global and sectoral aspects. Contribution of Working Group II to the Fifth Assessment Report of the Intergovernmental Panel on Climate Change,* ed. C.B. Field, V.R. Barros, D.J. Dokken, K.J. Mach, M.D. Mastrandrea, T.E. Bilir, M. Chatterjee, K.L. Ebi, Y.O. Estrada, R.C. Genova, B. Girma, E.S. Kissel, A.N. Levy, S. MacCracken, P.R. Mastrandrea, and L.L. White, 1–32. Cambridge: Cambridge University Press.

Kelly, P.M., and W.N. Adger. 2000. Theory and practice in assessing vulnerability to climate change and facilitating adaptation. *Climatic Change* 47: 325–352. https://doi.org/10.1023/A:1005627828199.

Kripalani, R.H., S. Inamdar, and N.A. Sontakke. 1996. Rainfall variability over Bangladesh and Nepal: comparison and connections with features over India. *International Journal of Climatology* 16: 689–703.

Kwarteng, A.Y., A.S. Dorvlo, and G.T. Vijaya Kumar. 2009. Analysis of a 27-year rainfall data (1977–2003) in the Sultanate of Oman. *International Journal of Climatology* 29: 605–617.

Mizina, S.V., J.B. Smith, E. Gossen, K.F. Spiecker, and S.L. Witkowski. 1999. An evaluation of adaptation options for climate change impacts on agriculture in Kazakhstan. *Mitigation and Adaptation Strategies for Global Change* 4: 25–41.

Pant, G.B., and K.R. Kumar. 1997. *Climates of South Asia.* Chichester, UK: John Wiley & Sons Ltd.

Piya, L., N. K. L. Maharjan, and P. Joshi. 2012. *Vulnerability of rural households to climate change and extremes: Analysis of Chepang households in the Mid-Hills of Nepal.* Selected Paper prepared for presentation at the International Association of Agricultural Economists (IAAE) Triennial Conference, Foz do Iguaçu, Brazil, 18–24 August, 2012.

Prescott-Allen, Robert. 2001. *The Wellbeing of Nations: A Country-By-Country Index of Quality of Life and the Environment.* Washington, DC: Island Press.

Pritchett, L., A. Suryahadi, and S. Sumarto. 2000. *Quantifying vulnerability to poverty: A proposed measure with application to Indonesia.* SMERU Working Paper 2437, Jakarta, Social Monitoring and Early Response Unit Research Institute (SMERU). http://documents.worldbank.org/curated/en/131651468751147913/pdf/multi-page.pdf. Accessed 25, Aug 2017.

Qin, N., J. Wang, G. Yang, X. Chen, H. Liang, and J. Zhang. 2015. Spatial and temporal variations of extreme precipitation and temperature events for the Southwest China in 1960–2009. *Geoenvironmental Disasters* 2 (4): 1–14. https://doi.org/10.1186/s40677-015-0014-9.

Raha, G., N.K. Bhattacharjee, M. Das, M. Dutta, and S. Bandyopadhyay. 2014. Statistical study of surface temperature and rainfall over four stations in north Bengal. *Mausam* 65 (2): 179–184.

Ravindranath, N.H., S. Rao, N. Sharma, M. Nair, R. Gopalakrishnan, A.S. Rao, S. Malaviya, R. Tiwari, A. Sagadevan, M. Munsi, N. Krishna, and G. Bala. 2011. Climate change vulnerability profiles for North East India. *Current Science* 101 (3): 384–394.

Rayner, S., and E.L. Malone, eds. 1998. *Human Choice and Climate Change. Volume Three: the Tools for Policy Analysis.* Columbus, OH, Battelle Press.

Sein, Z.M.M., and X. Zhi. 2016. Inter annual variability of summer monsoon rainfall over Myanmar. *Arabian Journal of Geoscience* 9: 469. https://doi.org/10.1007/s12517-016-2502-y.

Shahid, S. 2010. Rainfall variability and the trends of wet and dry periods in Bangladesh. *International Journal of Climatology* 30: 2299–2313.

Shukla, R., A. Chakraborty, and P.K. Joshi. 2017. Vulnerability of agro-ecological zones in India under the earth system climate model scenarios. *Mitigation and Adaptation Strategies for Global Change* 22: 399–425. https://doi.org/10.1007/s11027-015-9677-5.

Smit, B., and O. Pilifosova. 2001. *Adaptation to climate change in the context of sustainable development and equity.* ed. Impacts, adaptation and vulnerability contribution of working group II third assessment report of the intergovernmental panel on climate change. Cambridge: Cambridge University Press.

Soares, M.B., A.S. Gagnon, and R.M. Doherty. 2012. Conceptual elements of climate change vulnerability assessments: A review, February 2012. *International Journal of Climate Change Strategies and Management* 4 (1): 6–35. https://doi.org/10.1108/17568691211200191.

UNEP. 2002. *Global environment outlook 3: human vulnerability to environmental change.* United Nation Environmental Programme. London, Sterling, VA: Earthscan Publications Ltd.

Upgupta, S., J. Sharma, M. Jayaraman, V. Kumar, and N.H. Ravindranath. 2015. Climate change impact and vulnerability assessment of forests in the Indian Western Himalayan region: A case study of Himachal Pradesh, India. *Climate Risk Management* 10: 63–76.

Varikoden, H., K.K. Kumar, and C.A. Babu. 2013. Long trend trends of seasonal and monthly rainfall in different intensity ranges over Indian subcontinent. *Mausam* 64 (3): 481–488.

Wester, P., A. Mishra, A. Mukherji, and A.B. Shrestha, eds. 2019. *The Hindu Kush Himalaya Assessment—Mountains, Climate Change, Sustainability and People.* Switzerland AG, Cham: Springer Nature. ISBN 978-3-319-92288-1.

WMO. 2019 *World Meteorological Organization statement on the state of the Global Climate in 2018.* WMO-No. 1233, Switzerland.

World Bank. 2001. *A revised forest strategy for the World Bank Group.* Washington, DC: World Bank.

Yusuf, A. A., and H. Francisco. 2010. *Hotspots! Mapping Climate Change Vulnerability in Southeast Asia.* Economy and Environment Program for Southeast Asia. Singapore.

Chapter 2
The Bengal Duars: A Foothill Landscape of the Eastern Himalaya

Abstract The Bengal Duars region is located in the foothills of the Eastern Himalaya and characterized by unique physiography, climate, and rich biodiversity. Alongside, the region is also a cultural hub of indigenous people. The region has experienced a long evolutionary history and environmental change from the pre-colonial to the Independence period. The rules and controls over this region specifically on forest resources have changed over time. In this chapter, an effort has been made to explore the identity and evolution of the region and to describe physical as well as socio-economic characteristics of it.

Keywords Geographical region · Foothill landscape · Eastern Himalaya · Environmental change

2.1 Evolution of Bengal Duars and Its Spatial Identity

In the literature, Duars is well known as 'Doors' or 'Dooars'. This region has long historical evolution from the pre-Colonial to the post-Colonial period. Prior to the British annexation, the region was under the rule of Bhutanese and before that it was likely to belong to the Cooch Behar kingdom (Karlsson 2013). As Mughals had no interest in Dooars, they never conquered this region and it remained as a natural boundary or ecological border to restrict expansion of agriculture by the Mughals (Bhadra 1983; Karlsson 2013). In 1773, Cooch Behar had declared as a Princely state under the British East India Company. Dispute and negotiation between the Bhutanese in the hills and the Cooch Behar state on who actually owned what parts of Duars were a political issue (Sinha 1991). The East India Company had an interest in developing friendly relations with the Bhutan kingdom to open up a trade route to Tibet through Bhutan. Therefore, several mild steps were taken by the British even after continuous violations by the Bhutanese army (Collister 1996). In 1863, the British Company sent Ashley Eden on a diplomatic mission to Bhutan as the last attempt to settle the situation, but the mission was not successful. Finally, Eden was forced to sign the documents stating that both the Bengal and Assam Duars were Bhutanese territory and led to the Anglo-Bhutanese war in 1864–1865 (Karlsson 2013).

After the Anglo-Bhutanese war in 1864–1865, the Bengal Duars were permanently annexed by the Britishers. The boundary separating Duars from Bhutan hills was finally demarcated; thereafter, several districts were formed since 1867 (almost retaining the same shape). Dalingkot area (Kalimpong) was added from Bhutan to Darjeeling district in 1867 (Biswas 1999) and Jalpaiguri district was formed in 1869 consisting of three areas, namely Boda, Patgram, and Purbhbhag in the south-east, Baikunthapur in the north-west, and the Bengal Dooars in the north-east (Ray 2013). As noted in Rennie's book 'Bhotan and the story of the Dollar War' (Rennie 1866), 'Duars' was located along the foothill of the Himalayas, run along the whole length of Bhutan, from the river Teesta to the Dhunseri river. 'The strip of land with varying width from 10 to 30 miles, situated at foothills of the Himalaya, not properly belonging to Bhutan, wasted from Mughal and separates British frontier to subordinate chain of Bhutan hill. This tract was used for communication between plain and entire Bhutan through a series of mountain passes, locally known as Dooars' (Rennie 1866). As a gateway or entrance of Bhutan, the region was also known as 'Bhutanese Dooars'. On the 'sketch map of Bhotan and of the Dooars' by Rennie (1866), a track between the river Teesta and Manas was named as 'Bengal Dooars' (Fig. 2.1).

After Independence (1947), Darjiling (or Darjeeling) was merged with the state of West Bengal, comprising of hilly towns of Darjiling, Kalimpong, Kurseong, and the Terai areas of Siliguri. In 1869, Jalpaiguri district was formed and Cooch Behar District was evolved in 1950. The narrow strip of land, with a breadth of 20–30 miles and about 180 miles in length, assemblage with forest cover stretching between the river Sankosh in the east and the river Teesta in the west, and Cooch Behar on the south, forms the northern boundary of West Bengal is known as Bengal Duars or Western Duars (Fig. 2.2) and the other part in Assam District is popularized as

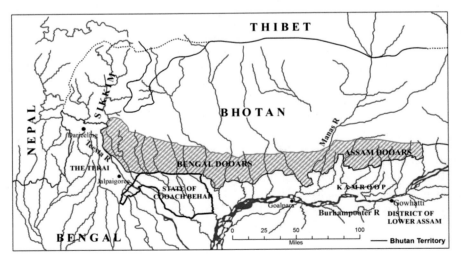

Fig. 2.1 Delineation of Bengal Duars region in the nineteenth century (Modified after Rennie 1866)

Assam Duars or Eastern Duars (Gruning 1911). Thus, river Sankosh acts as a divider of today's Bengal and Assam Duars and it was noted through the distribution of the rivers. In 2014, Jalpaiguri district was bifurcated and Alipurduar subdivision was formed as the new 20th district of the state and named as Alipurduar district. Kalimpong district was formed in 2017 after bifurcating Darjiling district. Presently, from the east of the river Teesta to the Sankosh river sharing the boundary of Bhutan in the north, assemblage with a rich forest cover, district Kalimpong, Jalpaiguri, Alipurduar and a small portion of Koch Bihar is popularly known as the Bengal Duars (or Dooars) region of West Bengal (Fig. 2.2).

2.2 Physiography and Drainage

The Bengal Duars lie in the foothills of Darjiling and Sikkim Himalayas with varieties of landforms. Broadly, this region can be divided into three physiographic units: the hills, the piedmonts, and the plain. The hilly region covers mostly Kalimpong district and the northern part of Jalpaiguri and Alipurduar district with convex ridges and deep valleys. The undulating topography and river terraces are those tropical features found in piedmont zone (Fig. 2.3). The alluvial fans of several rivers like Teesta, Torsa, Jaldhaka, Raidak, and Sankosh are observed in this intermediate zone and the alluvial plain has experienced frequent flood during the monsoon season. The Bengal Duars is a pivot of numerous rivers and rivulets, originating and passing through this region. The overall drainage pattern of the rivers is braided type due to the heavy load of sediments. The rivers are also controlled by structure, a large number of faults and lineaments are present in this region (Jana 1997). Sedimentation, shifting of river course, heavy rainfall and erosion in upper catchment cause flash flood mainly in Jalpaiguri, Alipurduar, and Koch Bihar districts every year (Sanyal 1969; Mukhopadhyay 1982).

2.3 Climate

It is the highest rainfall region in West Bengal (above 3000 mm/year), specifically during the monsoon season it experiences around 1800 mm rainfall. The mean annual rainfall varies from 2000 to 3500 mm. Thus, the climate is humid, marshy and gain an evil reputation for Malaria and Blackwater fever. The rainfall follows a typical monsoon pattern; the rainy season is from mid-May to mid-September and then there is a prolonged dry season. The amount of rainfall varies due to diverse physiography and marginally decreases towards the flood plains. Temperature (maximum and minimum) varies at different meteorological stations. Generally, March to June are the hottest months all over the region and maximum temperature ranges between 22 °C and 35 °C. Minimum temperature is observed during December and January and it varies from 2 °C to 15 °C.

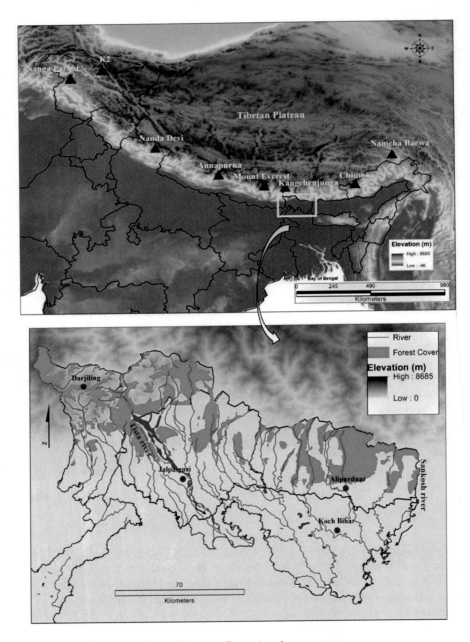

Fig. 2.2 Spatial identity of Bengal Duars (or Dooars) region at present

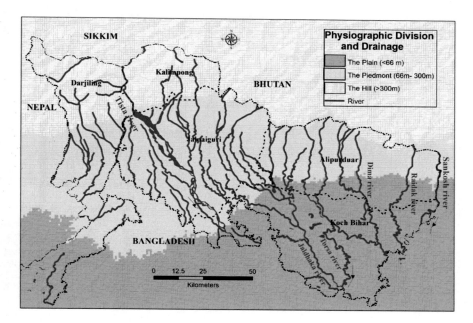

Fig. 2.3 Eastern Himalayan Foothills: a hub of rivers and rivulets (Source: SRTM data)

2.4 Soil

Mainly two kind of soils namely zonal and azonal are found in this region. The north hilly area is characterized by zonal soil and alluvial soil in plain comprises of azonal group. More specifically, over the Sub-Himalayan hilly ranges, the brown and submontane forest soil is generally found. The soil is acidic in nature and light in texture due to high rainfall and low temperature. The Piedmont region of foothills consists of boulder, gravel, pebble, sand with Tarai, and Bhabar soil is characterized by coarse texture, less porosity, and water retention capacity. The texture of soil changes towards the south, it became finer, alluvial in nature, and more acidic due to leaching in the flood plain of Koch Bihar (Fig. 2.4).

2.5 Culture

This region is a hub of indigenous culture and traditions and consists of Rabha, Koch, Mech, Toto, Rajbanshi, and Drukpa communities. During the pre-Colonial and Colonial period, the Rajbanshi community was a major community but there have been several debates regarding their origin and ethnic identity. After the introduction of tea plantation in the region, labour immigrated from Chota Nagpur and

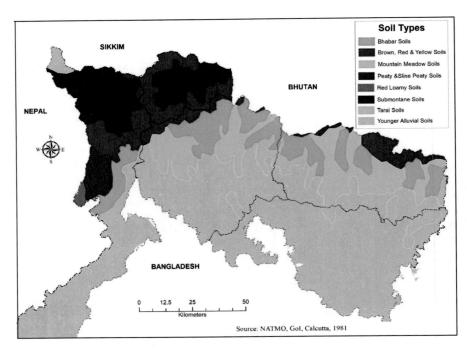

Fig. 2.4 Distribution of soil types in the northern part of West Bengal (Source: NATMO, GoI)

Santal Parganas, and as a result, other communities like Santal, Oraon, Mundas, and people from Bhutan all create a diverse cultural setting in this region. The area practices polyglot languages, a majority in the plain spoken Bengali language and Napali, Bhutanese, other local dialects are spoken in some hilly areas. However, with the progress of time and modernization, most of the indigenous cultures are losing their traditional values and practices. They are mostly residing in tea garden and forest villages in Terai and Duars.

2.6 Economy

Nearly 70% of the population living in the rural areas and the economy of this region is basically based on four 'T's like 'Tree', 'Tea', 'Timber' and 'Tourism'. Related to 'Tree' or 'Forest' and natural beauty of the landscape, some people are engaged in activities related to 'Tourism' in different National Parks, Biosphere Reserves, Sanctuary and worked as temporary labour under forest ranges. Otherwise, forest villagers are practicing agriculture and cultivate rice, vegetables, fruits, nuts, etc., and rearing common livestock. The socioeconomic status of the people is poor where the percentage of the working population ranges between 35 and 40% according to the 2011 Census of India data (Census of India 2011). However, 'Tea' and

'Timber' are the backbone of the economy for the commercialization of these resources. Many of the local people are engaged in these activities related to tea and timber industry but mostly they are found to be engaged in informal sectors as a worker and labour. In the forest sector, panchayats play an important role to generate employment through the implementation of social forestry, encouragement of self-help groups in the production of egg, milk, poultry, meat, and selling of local craft, utensils, etc.

References

Bhadra, G. 1983. Two Frontier Uprising in Mughal India. In *Subaltern Studies II*, ed. R. Guha. Delhi: Oxford University Press.

Biswas, K.R. 1999. [First print in 1918]. *A Summary of the Changes in the Jurisdiction of Districts in Bengal 1757–1916*. Calcutta: WBDG.

Census of India. 2011. http://www.censusindia.gov.in. Accessed on 05 July 2015.

Collister, P. 1996. [First print in 1987]. *Bhutan and the British*. New Delhi: UBSPD.

Gruning, J.F. 1911. *Jalpaiguri, Eastern Bengal and Assam District Gazetteers*. Govt. of West Bengal: WBDG.

Jana, M.M. 1997. Management and development of river basins in North Bengal using remote sensing techniques. *Photonirvachak, Journal of Indian Society of Remote Sensing* 25 (2): 105–111.

Karlsson, B.G. 2013. *Contested Belonging; An Indigenous People's Struggle for Forest and Identity in Sub-Himalayan Bengal*. Paperback, London and New York: Routledge.

Mukhopadhyay, S.C. 1982. *The Tista Basin*, 1–308. Calcutta: KP Bagchi.

Rennie, S. 1866. *Bhutan and the Story of the Dooars War*, 1–409. London: J. Murray.

Ray, S. 2013. *Transformations on the Bengal Frontier: Jalpaiguri 1765–1948*, 19–42. London and New York: Routledge.

Sanyal, C.C. 1969. The flood of 1968. *Geographical Review of India*. 31 (1): 47–53.

Sinha, A.C. 1991. *Bhutan: Ethnic Identity and National Dilemma*. New Delhi: Reliance Publishing House.

Soil Map (NATMO, GoI). n.d. SWID map. http://wbwridd.gov.in/swid/map.html. Accessed on 01 Dec 2019.

SRTM Data. n.d. DIVA GIS. http://www.diva-gis.org/. Accessed on 16 Aug 2018.

Chapter 3
Forestry of Bengal Duars

Abstract Forest is the backbone of Bengal Duars with rich biodiversity. A set of
drivers is at play to cause forest degradation over time. Growing anthropogenic
activities like mining, quarrying, logging, encroachment, tourism, etc., are affecting
the region as well. Economic activities often deject ecological benefit of the forest
which can further cause degradation of the whole landscape in future.

Keywords Forest · Biodiversity · Degradation

3.1 Forest and Biodiversity

Forest of Bengal Duars resembles an umbrella of West Bengal performing both
productive and protective roles. The sub-Himalayan Bengal contains a fair amount
of forest cover but the distribution is not uniform. Before the formation of the new
districts in the sub-Himalayan Bengal, forest areas cover mainly Darjiling, Jalpaiguri
and Koch Behar districts (Fig. 3.1). The density of forest cover is more in Darjiling
(38.23%) followed by Jalpaiguri (28.75%) and Koch Bihar (1.68%), respectively
(SFR 2009-10).

The diverse nature of physiography, climate and soil are responsible for the
changing nature of flora and fauna over the region. Edaphic, biotic and climatic
climax all together dominates the growth of forest. As this region is situated in monsoon climatic zone, it experiences heavy rainfall with plenty of sunlight, and the soil
becomes congenial for forest growth. There are miscellaneous tree varieties and are
used both as timber and non-timber forest products. Trees like Sal, Simul, etc., provide non-timber products along with timber and are considered as important forest
resource of the region.

Champion and Seth, 1968 [cited in Rakshit 2003] have classified Indian forest on
the basis of variation of temperature and rainfall as follows (Table 3.1)

1. Sal Forest of the plain (SF)
2. Riverine Forest (RF)
3. Wet Mixed Sal and moist deciduous forest (WM)

K. Sam, N. Chakma, *Climate Change in the Forest of Bengal Duars*,
SpringerBriefs in Environmental Science, https://doi.org/10.1007/978-3-030-73866-2_3

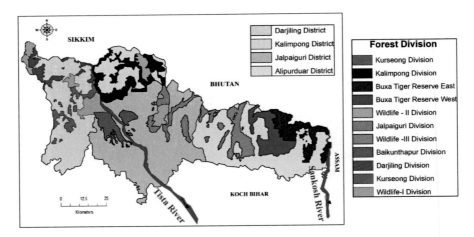

Fig. 3.1 Distribution of forest divisions in Bengal Duars

Table 3.1 Spatial detail about forest types and divisions

Forest types	Forest divisions	Name of the places/beats
Sal Forests of the plains (SF)	Wild life II, Baikunthapur, Jalpaiguri, Wild life III, BTR	Khunia, Dhupjhora, Buduram, Chilapata, Kodalbasti, Dumchi, Totapara, Sonakhili, Lataguri, Rajabhatkhawa, Damanpur, Dima
Riverine Forest (RF)	Wild life III, BTR, Jalpaiguri	Jaldapara, Nathua, Sulkapara, Balapara, Kumargram, Newland
Wet mixed Sal and Moist deciduous forest (WM)	Wild life II, Jalpaiguri, Wild life III, BTR	Murti, Gorumara, Chaprama, Kharbari, Rethi, Dalmore, Buxaroad, Rajabhatkhawa
Middle Hill forest (MH)	Kalimpong, BTR	Mo, Samsing, Buxaduar
Lower Hill forest (LH)	Kalimpong, Wild life III, BTR	Monpong, Noam, Lankapara, Buxaduar
Wet Temperature montane forest (WT)	Kalimpong	Lava, Lulagaon, Kalimpong
Alpine Forest (AF)	Kalimpong	Tangta

Sources: Rakshit 2003; Mallick 2010 & Forest Department, Government of West Bengal

4. Middle Hill forest (MH)
5. Lower Hill forest (LH)
6. Wet Temperature montane forest (WT)
7. Alpine Forest (AF).

The climatic and physiographic diversity has resulted in the appearance of a wide variety of birds, reptiles and animals. The wild elephant, buffalo, bison and rhinoceros are mostly found in this region. The red panda and brown bear are found in the hilly areas of Darjiling and Kalimpong districts. Among the carnivores, the Bengal tiger, bison, leopard, fishing cat, wolf, fox and jackal are also found (Fig. 3.2). Pig,

Fig. 3.2 Rhinoceros (**a**), Wild peacock (**b**), Elephant (**c**) and Bison (**d**) at Jaldapara National Park

poisonous snakes like python, cobra, pit viper, krait are common in this region. Beside these, black necked crane and pied hornbill are also found. The Trans-Himalayan migratory birds are the common visitors in the surrounding rivers, ponds and lakes in Terai and Duars. To protect all these wildlife and rich biodiversity, around one-third of the area has been assigned under the protected network. There are one Tiger Reserve, three National Parks and five Wild Life Sanctuaries located in the four districts of North Bengal covering Duars region.

3.2 Conceptualizing Drivers of Change

Bengal Duars region is well recognized for its valuable forest resources and biodiversity. This area has experienced a long-term historical transformation from the pre-British to the British and during the Independence periods. As stated, before the arrival of the British, this region was under Bhutanese rule and before that they most likely belonged to the Cooch Behar kingdom. Oral history claims that Duars was never a subject to the Mughals. As the forest tract was not an area of interest by the Mughals, perhaps it remained as a sparsely populated jungle tract (Karlsson 2013). The forest itself may have been a natural or ecological border for the expansion of the agricultural-based Mughals (Singh 1995).

During the British period, a drastic transformation has taken place through relatively open and previously natural (although not 'virgin') jungles into a controlled and man-made landscape (Karlsson 2013). Earlier, like other parts of Bengal, forest in Duars was under the revenue department, open for 'indiscriminate filling'. After 1857, with the creation of the Forest department, forest tracts were declared as Reserved (Karlsson 2013). The primary focus was on extraction and plantation of particular 'valuable trees to generate government revenue and also to satisfy the demand of timber for the construction of railways in India (Sivaramakrishnan 1995; Gadgil and Guha 1992). As a result, tea plantation was introduced in Tarai and Duars as well as sal tree (*Shorea robusta*) and teak tree (*Tectona grandis*) were planted mainly in Duars region. In order to establish monopoly, besides the need for timber, there was also a need to control unruly tribal groups in marginal and hilly tracts (Grove 1995). *Jhum* or temporary cultivation was prohibited there, but it was allowed in some areas (Progress Report of Forest Administration 1869). After a hard struggle of the Forest department to control forest fire in Reserved forest, a new problem was raised—the sal forest stopped to regenerate itself. The reduction of forest fire resulted in the tremendous spread of creepers and undergrowth, sal plant was stopped to grow (Karlsson 2013).

After this backdrop a radical solution was found by the British forester, a new method was introduced from the practice of shifting cultivation rather than scientific knowledge, and was named as '*Taungya*' revolution in Duars. After the introduction of *taungya* system in Duars, there was a necessity to develop forest villages because a skilled labour force who knew the technique of slash and burn cultivation was urgently required. Hence, indigenous shifting cultivators like Rava, Mech or Garo communities were back into the newly established forest villages whom once upon a time the British forester threw out from the forest (Jha 2010). During this period driver which was responsible for the exploitation of the forest resources include forest policy, introduction of tea plantation, wood extraction and logging. At that time population number was not so an effective factor behind changes in the forested landscape.

Conflicts over forest have continued from British period to the present, but a new 'ecological' dimension has been added to this. As Gadgil and Guha (1992) have pointed out the emergence of Wildlife Sanctuaries, National Parks and so-called Protected areas are largely new conflicting phenomenon in the post-Colonial period. To save the remaining forest, wildlife reserves, eco development projects and forest protection committees are developed. Since the late 1970s, large scale deforestation took place and during the 1990s number of felling of trees decreased substantially (Karlsson 2013).

Recent phase of transformation carried out beyond shifting cultivation through illegal felling, encroachment, mining (Figs. 3.3 and 3.4) and quarrying activities which further are enhancing flood vulnerability and decrease of vegetation stabilized gravel bars. Riverside boulder quarrying (Fig. 3.5) and deforestation are causing the shifting of river courses from straight to the meandering and braided pattern (Prokop and Sarkar 2012). Frequently reported illegal timber extraction, poaching of animal in local media and press reflects relatively poor enforcement and

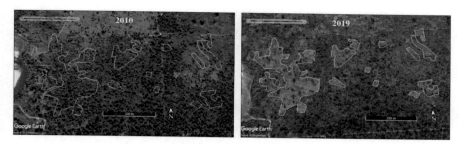

Fig. 3.3   Spatial view of fragmented patches within forest

Fig. 3.4   A clear cut tree felling observed at New Land Basti, Alipurduar

protection level that rises threat to the biodiversity hotspot. Although recent transformation occurs silently, it may cause devastating result in the whole landscape.
Anthropogenic activities mainly illegal felling and lack of forest protection measures are the responsible factors for changing of the forest ecosystem in the region.

3.3   Ecological Importance

Forest initiate life into the lifeless rock to make it a productive ecosystem. They
purify the air, water, moderate climate, maintain soil fertility and provide fuel, food,
medicine to humans and others. As being located in the foothills of the Eastern

Fig. 3.5 The activity of boulder quarrying along the river bed of Jayanti

Himalaya, the Bengal Duars region is prone to erosion accelerated by numerous river and rivulets. The ecological usefulness of forest in this region has been observed in their beneficial effects on river catchments, where they act as a regulator on streamflow, protect soils and prevent silting of dams, rivers and canals (Bhutia 1999; Sarkar 1989, 2000). Forest indeed are a blessing for the region and provide almost everything that people may require. The forest villagers used to collect fuelwood for cooking, fodder to practice animal husbandry and other non-timber products. Furthermore, forests are the backbone of foothill landscape and act as a protector from floods, landslides and soil erosion. Forests are sources of human needs for the tiny forest villagers. Besides the economic importance of forest, the growing awareness about the ecological cost of deforestation is also vital for the sustenance of the landscape of Bengal Duars.

References

Bhutia, P.T. 1999. *Environmental Degradation: Problems and Prospects—A Study in Kurseong sub-division of Darjeeling Himalaya*. Unpublished Ph.D. Thesis. University of North Bengal.
Gadgil, G., and R. Guha. 1992. *This Fissured Land; An Ecological History of India*, 120–237. Delhi: Oxford University Press.
Grove, R.H. 1995. *Green Imperialism. Colonial expansion, tropical island Edens and the origins of environmentalism, 1600-1860*. Delhi: Oxford University Press.
Jha, S. 2010. The struggle for democratizing forests: The forest rights movement in North Bengal, India. *Social Movement Studies* 9 (4): 469–474.

Karlsson, B.G. 2013. *Contested belonging; An indigenous people's struggle for forest and identity in Sub-Himalayan Bengal*. Paperback, London and New York: Routledge.

Mallick, J.K. 2010. Past and present status of the Indian Tiger in northern West Bengal, India: An overview. *Journal of Threatened Taxa* 2 (3): 739–752.

Progress Report of Forest Administration. 1869. *Progress report of forest administration in the lower provinces of Bengal during the year 1868-69*.

Prokop, P., and S. Sarkar. 2012. Natural and human impact on land use change of the Sikkimese-Bhutanese Himalayan piedmont, India. *Quaestiones Geographicae* 31 (3): 63–75. https://doi.org/10.2478/v10117-012-0010-z.

Rakshit, S.K. 2003. *Forest Resource: It's problem and prospects; a study on Darjeeling and Jalpaiguri District, West Bengal*. Ph.D. Thesis, North Bengal University, Darjeeling.

Sarkar, S. 1989. *Some consideration on soil eroaon hazardin Daqeelnig district (West of river Tista), W.B. Geographical Thought*, 45–54.

———. 2000. *Recent flood in sub-Himalayan North Bengal-causes and remedial measures—Bipanna Paribesh*, 288–293.

SFR. 2009-10. *State Forest Report, West Bengal*. Directorate of Forests, Kolkata: Government of West Bengal.

Singh, C. 1995. Forest, pastoralist and agrarian society in Mughal India. In *Nature, Culture, Imperialism*, ed. D. Arnold and R. Guha. Delhi: Oxford University Press.

Sivaramakrishnan, K. 1995. Colonialism and forestry in India: Imagining the past in the present politics, in comparative study of society and history. *Comparative Studies in Society and History* 37 (1): 3–40.

Chapter 4
Climate Change in Bengal Duars

Abstract Climate change is a natural phenomenon, but the present rate of change is quite faster than it was perceived before. In the era of the Anthropocene, human modification of the surrounding natural environment acts as an incentive to alter and change the rhythm of a natural system like climate. The present chapter is aimed not only to explore past and present changes of climatic variables but also to project future trends that may affect the Bengal Duars region.

Keywords Climate change · Natural phenomenon · Anthropocene

4.1 Trends of Climatic Variables

According to the IPCC 4th assessment report, climate change is a long-term change in climatic variables like temperature, precipitation, humidity, etc., and climate is defined as a statistics of 30 or 35 years of weather conditions. Therefore, to find out the change of climate, it is important to analyse the trends of climatic parameters. The study of climate change is now gaining importance in the scientific field of research, and scientists are recognizing the inherent variability of climate based on trend analysis (Meehl et al. 2000; Mantona et al. 2001; Pant and Kumar 1997; Rajendran et al. 2013; Radhakrishnan et al. 2017). From the old literature (Hamilton 1838), it appears that the temperature of the Terai and Duars never exceeded 27 °C. Now, the temperature has increased gradually beyond this limit. This region receives maximum rainfall (above 18,000 mm) during monsoon season and the trend of extreme rainfall events makes an understanding of the probability of flood incidence. Analysis has been done on the basis of daily and monthly gridded climatic data of rainfall (1901–2017) and temperature (1951–2017) dataset of India Meteorological Department (IMD), Government of India, and from the web link www.indiawaterportal.org/met_data/ another set of temperature data (1901–1950) has also been collected and analysed. Parametric and non-parametric trend tests with some ancillary climatic diagrams have been executed for the better understanding of detailed changes in the climate of the Bengal Duars region.

4.1.1   Temperature

The Mann-Kendall (Z) test and Sen's slope (Q) analysis revealed significant positive trends of maximum and minimum temperature for all the seasons with different
degrees of variability in the studied region. During the period of post-Monsoon
(October–December) and Winter (January-February) seasons, the rate of change of
temperature as well as deviation from normality are found to be subsequently high
for both the seasons (Figs. 4.1 and 4.2). The shift of mean temperature towards
warm seasons in the last 30 years increased the length and frequency of warm periods (Sam and Chakma 2019a). In case of temperature, changes are more sensitive
for minimum value. In the last two to three decades (1991–2017), the deviation of
minimum temperature is found to be high in comparison with the previous decades
(Fig. 4.3).

4.1.2   Rainfall

The region receives maximum rainfall during Monsoon season. In 1901–2017, a
significant positive trend has been observed not only during Monsoon season but
also in pre-Monsoon season (Fig. 4.4). Over the last two decades, dichotomy was
observed in rainfall pattern for some months during the Monsoon season and
extreme events became much more prominent in Monsoon and post-Monsoon
periods (Sam and Chakma 2019a). Specifically, in the months of April and May, a
significant positive trend of rainfall has been noted (Sam and Chakma 2019b).

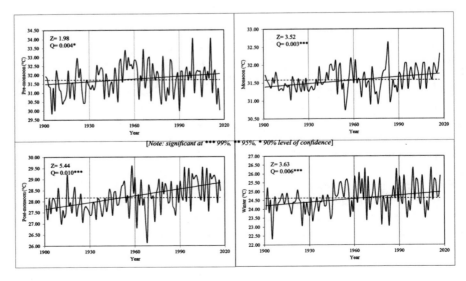

Fig. 4.1 Seasonal trend of maximum temperature

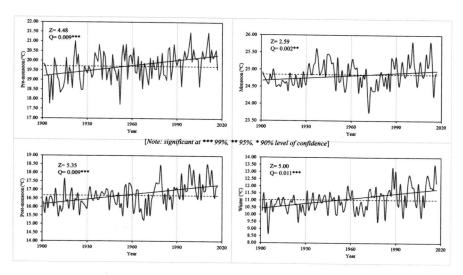

Fig. 4.2 Seasonal trend of minimum temperature

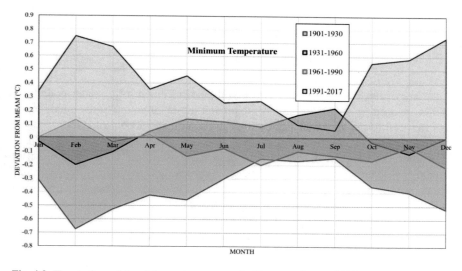

Fig. 4.3 Trend of monthly minimum temperature in 30 years subsets of 1901–2017

In rainfall analysis, the assessment of extreme events plays vital role in this flood-prone region. Analysis of multi-decadal oscillations is important to identify the relative difference between sub-period quantile and full-period quartile (Ntegeka and Willems 2008; and Willems 2013). Observations of multi-decadal oscillations of extreme rainfall with higher quantiles have been detected after the 1970s dissimilar accordance to the circular variation of anomalies (Sam and Chakma 2019a).

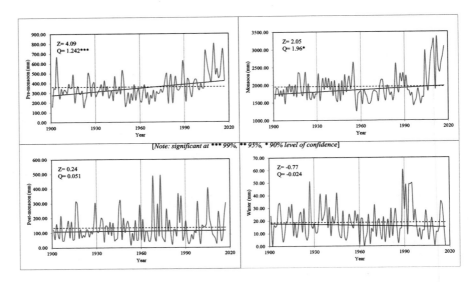

Fig. 4.4 Seasonal trend of rainfall

4.2   Projection of Future Climate

In the era of twenty-first century, through the advancement of climate modelling, GCMs are used by several researchers to predict future climate. Those climate models often fail to capture finer details of regional climatic scenarios due to coarser resolution of CMIP5 (~340 km²), specially to predict South Asian Monsoon climate (Rajendran et al. 2013). In such circumstances, the use of a high-resolution model or analysis of previous trends of climatic variables is necessary to predict future climate. In the present study, the future climate of Bengal Duars has been projected on the basis of the statistical analysis of the previous trends.

After evaluating more than 100 years' data of climatic variables, it has been observed that there is an oscillatory pattern of changes in temperature and rainfall. Since the 1990s, an identical change has been recognized for both annual average temperature and extreme rainfall (Figs. 4.5a and b). Moreover, the average temperature has increased at the rate of 0.023 °C⁻¹. Therefore, it may be inferred that in future decades average temperature will be increased at 0.2 °C or more than that (Fig. 4.5a). By following the pattern of extreme rainfall, the positive phase of oscillation from the 1970s onwards indicates that there is a possibility of 5% higher intensified extreme events in the next future decades (Fig 4.5b). Those future prediction of temperature and rainfall extremes can help further to the engineer and planner in decision making process by estimating the future climatic risks.

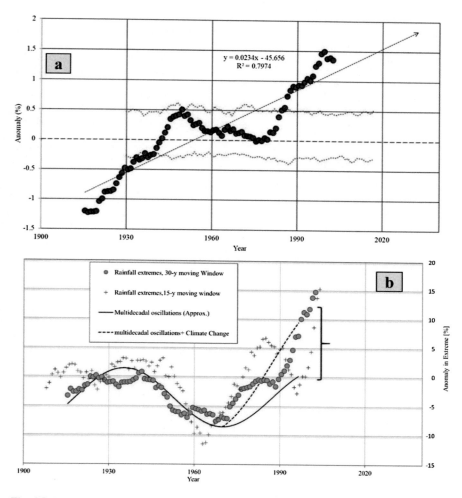

Fig. 4.5 Multi-decadal oscillations: (**a**) Average temperature and (**b**) Rainfall (1901–2017)

References

Hamilton, F. B. 1838. *The history, antiquities topography and statistics of Eastern India compris-ing the districts of Puraniya, Ronggopoor and Assam;* Collated from the Original Documents (i.e., from F Buchanan's Accounts of these districts 1807–1814 by Montgomery Martin), 3: London.

Mantona, M.J., P.M. Della-martab, M.R. Haylocka, K.J. Hennessyc, N. Nichollsa, L.E. Chambersa, D.A. Collinsb, G. Dawd, A. Finete, D. Gunawanf, K. Inapeg, H. Isobeh, T.S. Kestini, P. Lefalej, C.H. Leyuk, T. Lwinl, L. Maitrepierrem, N. Ouprasitwongn, C.M. Pagec, J. Pahalado, N. Plummerb, M.J. Salingerd, R. Suppiahc, V.L. Tranp, B. Trewinb, I. Tibigq, and D. Yee. 2001. Trends in extreme daily rainfall and temperature in Southeast Asia and the South Pacific: 1961–1998. *International Journal of Climatology* 21: 269–284. https://doi.org/10.1002/joc.610.

Meehl, G.A., F. Zwiers, J. Evans, T. Knutson, L. Mearns, and P. Whetton. 2000. Trends in extreme weather and climate events: issues related to modeling extremes in projections of future climate change. *Bulletin of the American Meteorological Society.* 87 (3): 427–436.

Ntegeka, V., and P. Willems. 2008. Trends and multidecadal oscillations in rainfall extremes, based on a more than 100 years' time series of 10 minutes' rainfall intensities at Uccle, Belgium. *Water Resource. Research* 44: 1–15.

Pant, G.B., and K.R. Kumar. 1997. *Climates of South Asia.* Chichester: John Wiley & Sons Ltd.

Radhakrishnan, K., I. Sivaraman, S.K. Jena, S. Sarkar, and S. Adhikari. 2017. A Climate Trend Analysis of Temperature and Rainfall in India. *Climate Change and Environmental Sustainability* 5 (2): 146–153.

Rajendran, K., S. Sajani, C.B. Jayasankar, and A. Kitoh. 2013. How dependent is climate change projection of Indian summer monsoon rainfall and extreme events on model resolution? *Current Science* 104: 1409–1418.

Sam, K., and N. Chakma. 2019a. An exposition into the changing climate of Bengal Duars through the analysis of more than 100 years' trend and climatic oscillations. *Journal of Earth System Science* 128 (67): 1–12. https://doi.org/10.1007/s12040-019-1107-8.

———. 2019b. Variability and trend detection of temperature and rainfall: A case study of Bengal Duars. *Mausam* 70 (4): 807–814.

Willems, P. 2013. Multidecadal oscillatory behaviour of rainfall extremes in Europe. *Climatic Change* 120: 931–944.

Chapter 5
Assessment of Vulnerability

Abstract Vulnerability-based approach includes both climatic and non-climatic factors under the domain of sensitivity, exposure and adaptive capacity. Assessment of vulnerability at spatial level is an utmost necessity in climate change-related impact studies as it helps the government authority to prepare sector-wise plans and to make aware local people about the situation so that they may adapt well with the changing condition of climate and associated risks. Therefore, the aim of this chapter is to estimate the level of vulnerability of the forest villages located in different forest divisions of the Bengal Duars region.

Keywords Vulnerability · Climate change · Risk · Forest village

5.1 Spatial Vulnerability

Assessment of vulnerability is regarded as a major step towards effective management of disasters and reducing risk (Brikmman 2006). With the passage of time, the concept of vulnerability has changed and is recognized as scale-dependent and multidimensional. Moreover, the effective utilization of Geographic Information System (GIS) tools in recent time facilitate the spatial dimension of vulnerability by including different factors to understand root causes of vulnerability throughout the extent (Montz and Evans 2001; Roy and Blaschke 2015; Lim et al. 2016). UNDP (2004) defines vulnerability as 'a human status or operation resulting from physical, societal, economic, and environmental factors, which influence the livelihood and scale of impairment of the impact of a given hazard'. As climate change is an inescapable phenomenon and a global issue, therefore, it is a novel addition in the vulnerability study. IPCC (2001) considered vulnerability as a susceptibility of the system towards changes and ability to adjust or cope. It is a function of exposure, sensitivity and adaptive capacity. The assessment of vulnerability to climate change can provide information to identify natural and artificial controls that influence or guard against ripple effects of climate change which is occurring now or may happen in future (IPCC 2001). Therefore, vulnerability in such context is an expression of the relationship between climate change impacts and the system. A system will be more vulnerable if the impact of climate change is high and adaptability of a system is low (UNEP, 2002).

5.1.1 Method of Analysis

Village-level vulnerability analysis of forested landscape in Bengal Duars has been done by applying Analytical Hierarchy Process (AHP), where individual factors are generally prioritized according to the views of experts or users (Saaty and Vargas 2000). In the second-generation model of vulnerability assessment (Füssel and Klein 2006) other than climatic factors, non-climatic factors are also recognized as important drivers of a system to climate change. In this study, 21 factors (including climatic and non-climatic) are selected and assigned in three domains as exposure, sensitivity and adaptive capacity (Table 5.1). All factors are standardized before overlay analysis in the GIS environment. Besides, 1 km buffer around the forest boundary has been considered in the analysis as per MoEFCC guidelines to explore

Table 5.1 Detailed account of indicators used in the study

Domain	Indicator	Data source	Period
Exposure	1. Annual average rainfall	Directorate of Agriculture, Govt. of India; IMD, Govt. of India; CWC, Govt. of India; and Sarkar (2004)	1971–2002
	2. Annual average temperature		1971–2002
	3. Deviation in temperature		1930–2002
Sensitivity	4. Physiography	ASTER GDEM	2011
	5. Slope		
	6. Clay mineral index	Landsat 5 (USGS-GloVis)	2010
	7. Distance from river	Digitized feature from ASTER GDEM	2011
	8. Distance from roads	Survey of India Topographical sheet	1930s, 1970s
	9. Land cover	Landsat 5	2010
	10. Change in land cover (forest cover)	Landsat 5	1990, 2010
	11. Population growth	Census of India	2011
	12. Population density		
	13. Percentage of non-workers		
Adaptive capacity	14. Number of health institutions	Census of India	2011
	15. Literacy rate		
	16. Access to district roads		
	17. Availability of cooperative/ commercial bank		
	18. Availability of regular mandis/ markets		
	19. Availability of tap water		
	20. Power facility		
	21. Access to mobile		

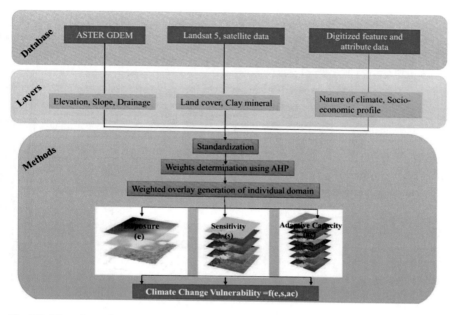

Fig. 5.1   Flow chart of methodology used in the study [Sam and Chakma 2018]

adjacent disturbances and influences at the forest edge. The detailed framework of vulnerability assessment is specified in Fig. 5.1.

5.1.2   Exposure

Exposure defines the nature and the extent of change of climatic variables like temperature, rainfall, extreme weather events, etc. (Brenkert and Malone 2005). In this analysis, variables like temperature and its deviation, and rainfall are considered. The forest divisions of Wildlife-II and Wildlife-III, Jalpaiguri and Baikunthapur are relatively more exposed to climatic stimuli (Fig. 5.2a). The annual average temperature is a relatively dominant determinant in comparison to the other two variables (Sam and Chakma 2018).

5.1.3   Sensitivity

The concept of sensitivity is alike immunity of a system, affected by climatic variability or change. It is an important structure of a system sensible towards changes. Overall, ten factors are carefully chosen including physical, demographic and socio-economic aspects to analyse the sensitivity of the landscape. Mostly, fringe areas of

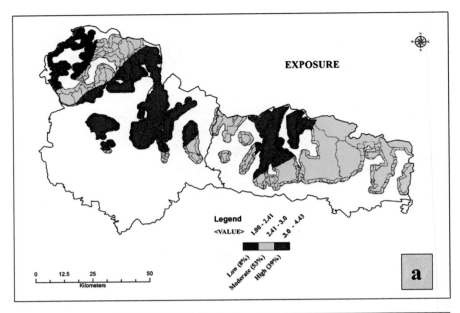

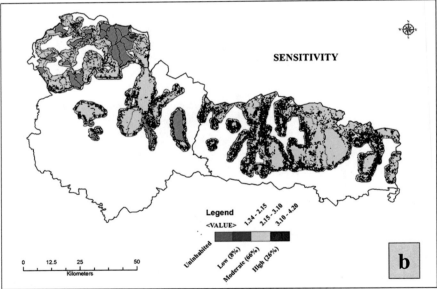

Fig. 5.2 Domains of vulnerability in Bengal Duars: (**a**) Exposure, (**b**) Sensitivity, (**c**) Potential impact and (**d**) Adaptive capacity [Sam and Chakma 2018]

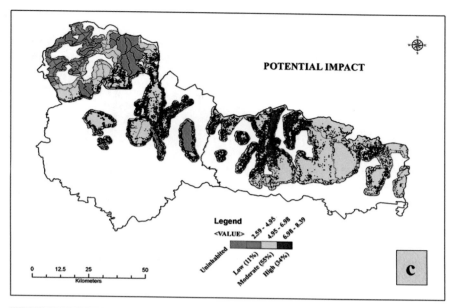

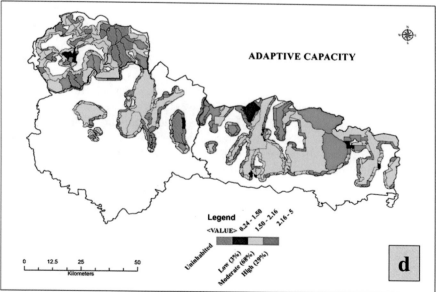

Fig. 5.2 (continued)

Wildlife-III, Buxa Tiger Reserve and Jalpaiguri divisions are found to be more sensitive due to anthropogenic disturbances that lead to changes in land cover (Fig. 5.2b). Exposure and sensitivity facets are executed to estimate potential impact, and it has been found that about 34% of the area has experienced high potential impacts of both climatic and non-climatic variables (Fig. 5.2c).

5.1.4 Adaptive Capacity

Adaptive capacity is an ability of a system to make adjustment or adapt to climate change or to cope with the consequences (IPCC 2001). The capacity of adaptation is regarded as a crucial property to reduce vulnerability (Adger et al. 2007; Williamson et al. 2012). In this domain of analysis, eight socio-economic and infrastructural variables are selected. Overall, about 68% area has been found as of moderate adaptive capacity and only 29% of the area experienced relatively high adaptive capacity due to better convenience of essential assets and amenities (Fig. 5.2d).

5.2 Discussion

Assessment of the level of vulnerability at local or micro level plays an important role to understand the nature and characteristics of risk factors at the local level which may guide the respective government authorities for adapting appropriate measures to reduce vulnerability. After the assessment of three domains of exposure, sensitivity and adaptive capacity, the overall index of vulnerability is alarming as it shows that around 61% of the area is highly vulnerable to climate change (Fig. 5.3a). Most of the villages are located in Buxa Tiger Reserve East, Baikunthapur, Wildlife-II and Wildlife-III forest divisions. On the basis of the projected change of temperature in Chap. 4 (Sect. 4.2), the level of vulnerability of 2040s has also been projected in Fig. 5.3b. Most of the area has been projected to be highly vulnerable to climate change in the 2040s and need to draw the attention of the respective government authorities to adapt appropriate strategies for the rural people of the Bengal Duars.

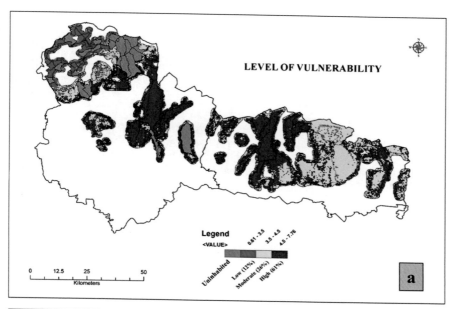

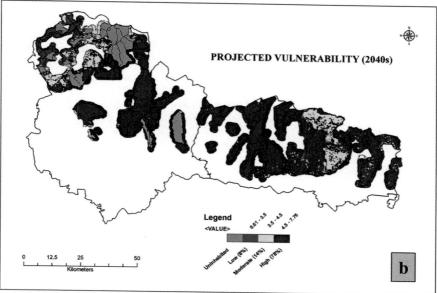

Fig. 5.3 Assessment of vulnerability in Bengal Duars: (**a**) Present scenario and (**b**) Projected scenario (2040s)

References

Adger, W.N., S. Agrawala, M.M.Q. Mirza, C. Conde, K. O'Brien, J. Pulhin, R. Pulwarty, B. Smit, and K. Takahashi. 2007. Assessment of adaptation practices, options, constraints and capacity. In *Climate Change 2007: impacts, adaptation and vulnerability. Contribution of working group II to the fourth assessment report of the intergovernmental panel on climate change*, ed. M.L. Parry, O.F. Canziani, J.P. Palutikof, P.J. van der Linden, and C.E. Hanson, 717–743. Cambridge: Cambridge University Press.

ASTER GDEM Scenes. 2011. ASTGTM2_N27E088, N27E089, N26E088, N26E089; as a product of METI and NASA. https://earthexplorer.usgs.gov/. Accessed 25 Feb 2017.

Brenkert, A.L., and E.L. Malone. 2005. Modeling vulnerability and resilience to climate change: a case study of India and Indian States. *Climatic Change* 72: 57–102. https://doi.org/10.1007/s10584-005-5930-3.

Brikmman, J. 2006. Measuring vulnerability to promote disaster-resilient societies: Conceptual frameworks and definitions. In *Measuring vulnerability to natural hazards- towards disaster-resilient societies*, ed. J. Brikmann, 9–54. New York: United Nations University.

Census of India. 2011. http://www.censusindia.gov.in. Accessed 10 Aug 2015

Füssel, H.M., and R.J.T. Klein. 2006. Climate change vulnerability assessments: an evolution of conceptual thinking. *Climatic Change* 75: 301–329. https://doi.org/10.1007/s10584-006-0329-3.

IPCC. 2001. *Climate Change 2001: Synthesis Report*. A contribution of working groups I, II, III, Third assessment report of the Intergovernmental Panel on Climate Change. Cambridge: Cambridge University Press.

Lim, S.J., H.J. Park, H.S. Kim, S.I. Park, S.S. Han, H.J. Kim, and S.H. Lee. 2016. Vulnerability assessment of forest landslide risk using GIS adaptation to climate change. *Forest Science and Technology* 12 (4): 207–213. https://doi.org/10.1080/21580103.2016.1189853.

Montz, B.E., and T.A. Evans. 2001. GIS and social vulnerability analysis. In *Coping with Flash Floods. NATO Science Series (Series 2. Environmental Security)*, ed. E. Gruntfest and J. Handmer, 77. Dordrecht: Springer.

Roy, D.C., and T. Blaschke. 2015. Spatial vulnerability assessment of floods in the coastal regions of Bangladesh. *Geomatics, Natural Hazards and Risk* 6 (1): 21–44.

Saaty, T.L., and L.G. Vargas. 2000. *Models, methods, concepts, and applications of the analytic hierarchy process*. Boston: Kluwer Academic Publishers.

Sam, K., and N. Chakma. 2018. Vulnerability profiles of forested landscape to climate change in Bengal Duars region, India. *Environmental Earth Sciences* 77 (459): 1–11. https://doi.org/10.1007/s12665-018-7649-2.

Sarkar, S. 2004. *Jalpaiguri Jelar Abhayar Sankal o Akal*. Jalpaiguri: Kirat Bhumi.

UNDP. 2004. *Reducing disaster risk: a challenge for development. A global report*. New York: Bureau for Crisis Prevention and Recovery (BRCP.

UNEP. 2002. *Global environment outlook 3: human vulnerability to environmental change*. United Nation Environmental Programme. London, Sterling, VA: Earthscan Publications Ltd.

USGS Global Visualization Viewer (GloVis). (n.d.). Landsat 5 data. http://glovis.usgs.gov. Accessed 20 Jan 2016.

Williamson, T., H. Hesseln, and M. Johnston. 2012. Adaptive capacity deficits and adaptive capacity of economic systems in climate change vulnerability assessment. *Forest Policy and Economics* 15: 160–166.

Chapter 6
Perception and Responses of the Forest Villagers

Abstract In the discourse of human–environment relationship, environmental change can be studied through human's perception and responses. In this section, climatic and non-climatic changes and its impact on life has been studied by considering people's perception, which is based on their experiences, belief and value of life, and responses that are reliant on action and reaction of life of forest villagers. Livelihood is an economic dimension of life and a means of securing basic needs of life and an interesting outcome of human–environment relationship. It can be an asset if the relationship is synchronised well, and if not, can create complex and crisis in livelihood.

Keywords Perception · Responses · Life · Livelihood

6.1 People's Perception About Environmental Changes

The forest in the foothill landscape of Eastern Himalaya had experienced enormous changes from pre-British to after Independence. During 1990–2010, the maximum transformation has taken place from dense forest to open forest cover specifically in Alipurduar district and the recent phase of disturbances are stimulated by felling of trees, encroachment, mining and quarrying along the river bed (Prokop and Sarkar 2012; Sam and Chakma 2021). There are numerous households settled in forests of the Bengal Duars region. The inhabitants are mostly indigenous and have experienced a long journey of struggle in such a landscape, and few of them have migrated from surrounding areas. In a way, they are the encyclopaedia of traditional knowledge and experience. As they are living with nature, the study of their perception about changes eventually will help to detect the problems, to recognize the adaptation strategies, and to formulate a better plan and climate-resilient system (Maraseni et al. 2005; Manandhar et al. 2012; Crona et al. 2013).

6.1.1 Survey Sketch and Methods

As this study aimed to assess people's perception and responses in most critical
zone of forest towards changes thus, based on the level of vulnerability (see Chap.
5) six forest villages were selected located in highly vulnerable zone (Fig. 6.1). The
survey was conducted between 2015 to 2018, and primary data was collected
through semi-structured questionnaire survey. The detailed profile of surveyed
households ($n = 196$) living in the selected six forest villages are given in Table 6.1.
Through the questionnaire survey, views of three generations of each family with
age ranges between 15 and 83 years were collected. The majority of the villagers
settled from 1948 to 1990. Most of the households are in BPL (Below Poverty
Level) category. Cronbach's alpha (Hair et al. 2006) has been used to judge the reliability of their perceptions and PCA (Principal Component Analysis) with cluster
analysis was executed to group them.

6.1.2 Perception About Changes in Forest

The forest has close semblance with the region. It protects the region like an
umbrella from certain disturbances. Forest villagers have observed lots of changes
occurring during their habitation here. It is interesting to know from the inhabitants

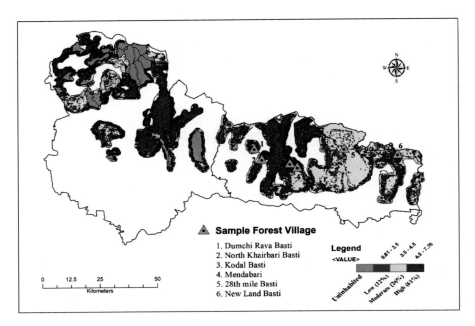

Fig. 6.1 Location of the selected forest villages in Bengal Duars

Table 6.1 Demographic and socio-economic characteristics of sample forest villages (n = 196)

Variables	Forest Villages													
	Dumchi Rava Basti		Uttar Khairbari Basti		Kodal Basti		Mendabari		28th Mile Basti		New Land Basti		F	Sig.
	Mean	SD	Mean	SD	Mean	SD	Mean	SD	Mean	SD	Mean	SD		
1. Household size	4.33	1.51	5.00	0.82	4.13	1.25	4.42	1.35	4.90	1.80	6.00	1.92	3.17	0.011*
2. Age of respondent	31.00	27.75	40.25	38.42	38.50	30.95	44.00	38.10	35.25	32.06	41.20	30.75	2.19	NS
3. Caste/category (S.C = 1, S.T = 2, O.B.C = 3, General = 4)	2.00	0.00	1.75	0.50	2.00	0.00	2.43	0.66	2.95	0.94	1.85	0.88	5.52	0.000**
4. Religion (Hindu = 1, Islam = 2, Buddhist = 3, Christian = 4)	4.00	0.00	3.25	1.50	1.38	1.06	1.31	0.94	1.50	0.89	1.20	0.62	13.09	0.000**
5. Year of resident before 1947 = 1, 1948–1990 = 2, after 1990 = 3	1.33	0.52	1.75	0.50	1.38	0.52	1.20	0.41	1.10	0.31	1.25	0.44	4.06	0.002**
6. Number of males	1.83	0.98	3.25	0.96	2.00	0.58	3.15	1.28	2.70	1.26	3.05	1.36	1.72	NS
7. Number of females	1.50	0.84	1.75	0.50	2.38	1.77	2.60	1.54	2.20	1.20	2.90	1.25	1.54	NS

(continued)

Table 6.1 (continued)

Variables	Forest Villages													
	Dumchi Rava Basti		Uttar Khairbari Basti		Kodal Basti		Mendabari		28th Mile Basti		New Land Basti		F	Sig.
	Mean	SD	Mean	SD	Mean	SD	Mean	SD	Mean	SD	Mean	SD		
8. Education level of male: primary = 1, secondary = 2, higher secondary = 3, graduation = 4	1.33	1.21	2.00	0.82	1.63	0.92	2.74	1.87	1.80	0.89	1.85	0.49	0.77	NS
9. Education level of female: primary = 1, secondary = 2, higher secondary = 3, graduation = 4	0.33	0.82	1.50	1.29	1.75	1.16	1.64	1.14	1.32	1.06	1.50	0.83	1.44	NS
10. Economic status: BPL = 1, APL = 2	1.33	0.52	1.25	0.50	1.25	0.46	1.20	0.40	1.20	0.41	1.40	0.50	0.59	NS
11. Livelihood options for male	2.00	0.63	2.25	0.96	1.71	0.76	1.79	0.76	1.55	0.69	1.80	0.70	0.89	NS
12. Livelihood options for female	1.83	0.41	1.25	0.50	1.50	0.53	0.87	0.53	0.68	0.58	0.60	0.50	7.73	0.000**

*Significant at 5% level ($P < 0.05$), **Significant at 1% level ($P < 0.01$), NS denotes no significant difference

Table 6.2 Test of reliability about changing forested landscape

Scale of measurement	Indicators	Cronbach's alpha if item deleted
3 = Increase/improve 2 = Decrease/degrade 1 = No idea	TS	0.94
	FS	0.85
	FC	0.93
	FH	0.87
	EN	0.85
	GR	0.75
	IP_TF	0.89
Overall Cronbach's reliability coefficient alfa (α) Acceptable limit of coefficient (α) is > 0.07		0.87
P value		0.001

Source: Field Survey

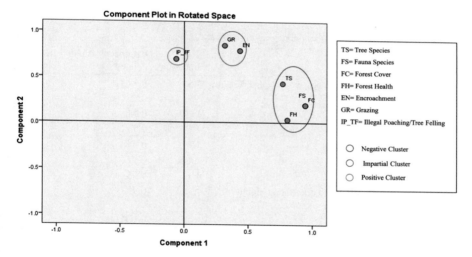

Fig. 6.2 Clustering of perceived changes in the forest

about the forest health and activities taking place in the area. Seven reliable variables are chosen for the study (Table 6.2). Forest villagers have identified positive changes in illegal activities such as poaching and tree felling. Impartial perception has been found in case of encroachment and grazing activities in forested landscape. The condition of forest cover, forest health, tree species and faunal species are deteriorating, and therefore found in a negative cluster (Fig. 6.2).

6.1.3   Perception About Change of Climate

The people's perception about the climate change has been collected based on belief topologies such as (1) climate is changing but the human has no contribution; (2) climate is changing and human has a contribution to climate change; (3) climate is

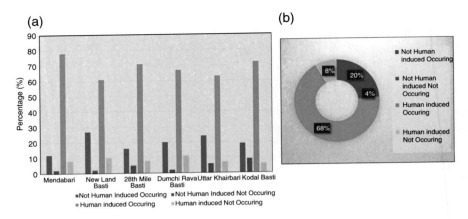

Fig. 6.3 Belief topology of climate change

Table 6.3 Test of reliability about changing climatic attributes

Scale of measurement	Indicators	Cronbach's alpha if item deleted
1 = No idea, 2 = Listen from others, 3 = Observed/felt by own	LS	0.97
	WS	0.95
	LCW	0.94
	OS	0.93
	ER	0.93
	OLR	0.97
	UF	0.96
Overall Cronbach's reliability coefficient alpha Acceptable limit of coefficient (α) is > 0.07		0.95
P value		0.00

Source: Field Survey

not changing but human has a contribution to climate change; and (4) climate is not changing and human has no contribution. Majority of the villagers believed that climate change is occurring and caused by human actions (68%). The topology of not human-induced and occurrence of climate change is about 20%, but few households were disagreeing with the occurrence of climate change either due to human induce or not (8% and 4%), see Fig. 6.3.

To examine the changes of climatic variables with reliable experiences (Table 6.3), villagers' perception have been collected and analysed in the present study. Through cluster analysis, three types of perceived groups have been identified. Villagers have a strong perception about changing behaviour of overlapping seasons, longer summer, less cool winter, extreme rainfall and late start of winter. On the other hand, unconventional fog formation and overall low rainfall have perceived less and warmer summer has perceived moderately by the households (Fig. 6.4).

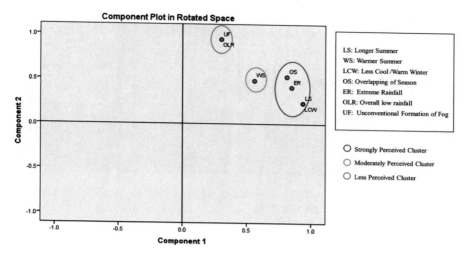

Fig. 6.4 Clustering of perceived changes in climatic attributes

6.2   Responses of Forest Villagers on Their Life and Livelihoods

The life and livelihoods of the inhabitants of the forest villages are often a struggle with various problems and consequences. Change in climatic variables are further enhancing troubles through the frequent incidence of extreme events, temperature abnormalities, overlapping of seasons and so on.

6.2.1   Struggle in Life

Most of the forest villages in Bengal Duars are located in remote areas surrounding forests and without good transport, communication and basic amenities. In some cases, dry river beds act as road and connecting path with seasonal gap in monsoon period (Fig. 6.5a). During the monsoon season, the situation becomes shoddy with bank-full river beds causing floods in the areas. Normally, villagers have to travel 5–15 km or more in order to avail better education, health and market facilities. During the monsoon season, forest villages remain cut off from the neighbouring areas and are compelled to struggle for their daily needs. During a field survey (2018) in Dumchi Rava Basti, an inhabitant responded that:

> During flood events, we are trapped within the house and had to spend several weeks without electricity, job opportunity and interaction with the nearby areas. There was fear of animal attack also.

Fig. 6.5 (**a**) Use of dry river bed for communication purposes; (**b**) Construction of wooden house above a certain height in the 28th Mile Basti

Human–animal conflict is a serious threat to the forest villagers in their daily life. Temporal changes in forest cover of the study areas results in loss of biodiversity, which directly affects food, shelter and breeding habitats of the wild animals. Fragmented wild habitats experience amplified vulnerability of wild life in terms of poaching, predation and invasion of exotic species into remaining forest habitats, and simultaneous encounter of humans with wild life results in loss of property, crops, assets and life. The houses in the forest villages have a unique style. These are made of wooden pillars and usually are constructed at 4–5 ft height above the ground to avoid flood and animal attacks (Fig. 6.5b).

6.2.2 Livelihood Dynamics

Forest village dwellers settled in Bengal Duars through a historical period and have experienced changes in socio-cultural and economic systems. With time, occupation of forest-dependent people have changed also with a direct impact on their livelihood. The studied region consists of several indigenous communities who are living here since the pre-Colonial period, for example, Mech, Garo and Toto. These people preferred to settle in isolated ecological zones. They live in a primitive form and practice shifting cultivation (Ray 2013). Mech are a nomadic community, who wandered from one place to another in the jungles and use *jhum* practices to cultivate cotton and paddy. During the British colonial period, natural and open jungles were transformed into human-controlled landscape. The main focus was on earning revenue through the extraction of valuable trees and construction of railways by the British rulers. For this, they also controlled primitive livelihood of people by restricting *jhum* cultivation (Karlsson 2013). Tea gardens were established in 1874–75, and large tracts of forest cover were eventually declared as wasteland. Initially, local people started working in the tea gardens as labourers but soon due to the requirement of heavy labour force in this labour-intensive industry large number of labourers were recruited from Chotanagpur, Bihar and adjacent areas populated by Oraon,

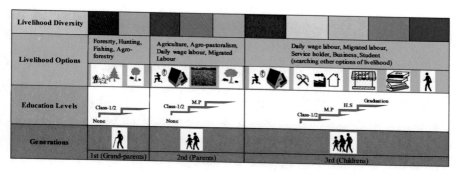

Fig. 6.6 Generation-wise changes in livelihood pattern and its diversity, Bengal Duars

Munda and Santal communities. During this land transformation period, agricultural settlers were also coming from the southern part of (Western) Duars. Thereafter, when the sal forest stopped regenerating itself, the *jhuming* system was reintroduced in the region in the name of *taungya* cultivation by the Forest Department itself. Requirement of indigenous community with their traditional knowledge about *jhum* or shifting cultivation occasioned the establishment of forest villages in 1915/1916, and the villagers took good care of the forest resources (Karlsson 2013).

Recent changes in people's livelihoods in the Bengal Duars has been analysed in the present study by amassing generation-wise occupational changes through a questionnaire survey during 2015–2018. It has been found that the first-generation (grandparents) members had little option of choices as means of livelihood, and therefore, were involved in agroforestry and hunting, and worked as forest labour (Fig. 6.6). The second-generation (parents) members were observed to be accustomed with diversified livelihood options. They were found to be engaged in agriculture, livestock rearing, daily wage labour and some of them even migrated to neighbouring foreign country of Bhutan, and others went to the far-off states like Kerala, Delhi etc. in search of better job opportunities. With the passage of time, more diversification in livelihoods have appeared. Presently, the young generation wants to be educated well so that they may have better job opportunity to get a better job as well as to enhance livelihood for their well-being. They also have different options of livelihoods likewise business, worked as a labour force within and outside the state. Nowadays, youths are becoming more educated to get permanent government or private jobs, and households affianced in small businesses are not uncommon too (Fig. 6.6).

6.2.3   Livelihood Complex and Crisis

Presently, the forest villagers in Bengal Duars are less dependent (~20%) on forest resources. Dependence on forest is mostly due to collection of fuelwood, forage and fodder purposes. Fuelwood is an essential substance in their daily life, and it is

Fig. 6.7 (**a**) Collection of fuelwood by a villager in 28th Mile Basti; (**b**) Storage of fuelwood in the house in 28th Mile Basti

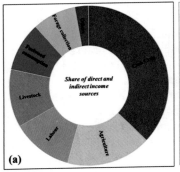
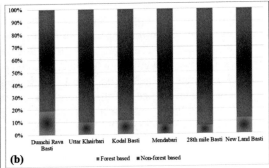

Fig. 6.8 (**a**) Income from different sources; (**b**) Share of dependency on forest and non-forest-based sectors

collected from forest (Fig. 6.7a). In the monsoon period (June–September), they have to face the distressing effects of flood and marshy environment. For this reason, they have to store fuelwood in advance before the monsoon period (Fig. 6.7b). Besides these, occasionally, they have to face harassment from the forest department authorities with allegations of illegal felling and poaching activities. Now people of the forest villages are mostly involved in non-forest-based activities like agriculture, cash crop production, livestock rearing, etc. (Fig. 6.8). When asked about the reason behind more dependency on non-forest-based activities, one villager replied that:

> *If the forest department starts plantation and other forest based activities on regular basis, and engage us regularly, then we don't need to migrate to other places in search of jobs.*

MGNREGA (Mahatma Gandhi National Rural Employment Guarantee Act), a central government act, has been implemented in these areas and the local residents are provided job cards to get job for 100 days in a year. But, the villagers expressed their unhappiness because they seldom get work and it is barely for 10–15 days/year on a Rs. 160–200/day (villager's opinion) basis during 2015–2018. As a result, a

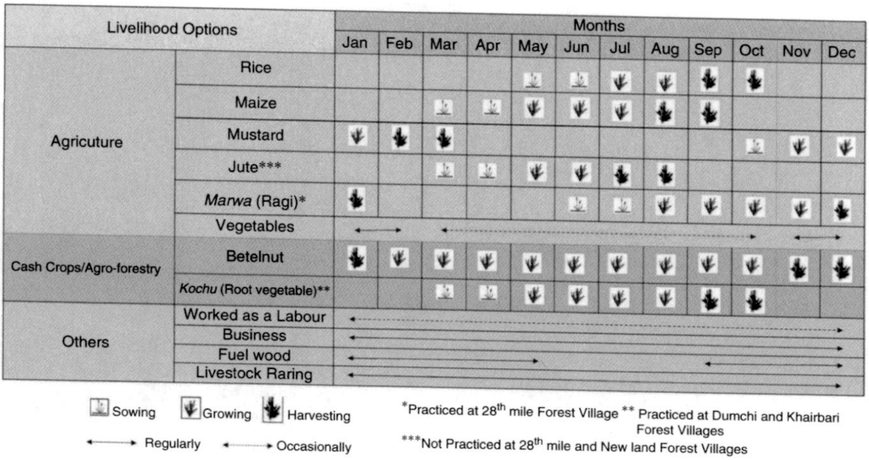

Livelihood Options		Months												
		Jan	Feb	Mar	Apr	May	Jun	Jul	Aug	Sep	Oct	Nov	Dec
Agriculture	Rice					Sowing	Sowing	Growing	Growing	Harvesting	Harvesting		
	Maize			Sowing	Sowing	Growing	Growing	Growing	Harvesting	Harvesting			
	Mustard	Growing	Harvesting	Harvesting							Sowing	Growing	Growing
	Jute***			Sowing	Sowing	Growing	Growing	Harvesting	Harvesting				
	Marwa (Ragi)*	Harvesting					Sowing	Sowing	Growing	Growing	Growing	Growing	Growing
	Vegetables	← Regularly →		← Occasionally →							← Occasionally →		
Cash Crops/Agro-forestry	Betelnut	Harvesting	Growing	Growing	Growing	Growing	Growing	Growing	Growing	Growing	Growing	Harvesting	Harvesting
	Kochu (Root vegetable)**			Sowing	Sowing	Growing	Growing	Growing	Growing	Harvesting	Harvesting		
Others	Worked as a Labour	← Regularly →											
	Business	← Regularly →											
	Fuel wood	← Regularly →					← Occasionally →			← Occasionally →			
	Livestock Raring	← Regularly →											

Sowing Growing Harvesting *Practiced at 28th mile Forest Village ** Practiced at Dumchi and Khairbari Forest Villages

←—→ Regularly ←--→ Occasionally ***Not Practiced at 28th mile and New land Forest Villages

Fig. 6.9 Seasonal calendar of livelihood practiced by forest villagers in Bengal Duars

large number of villagers have to migrate to other states for better job opportunities. Most of the people who migrated on seasonal basis are less educated (mostly educated up to primary level, i.e., class 4 only) and unskilled. Therefore, they do not get better jobs in other states too. Such outmigration occurs because of crisis of livelihoods in their own places. Villagers are engaged in agricultural activities also and cultivated crops like paddy, jute, maize, mustard and vegetables during the monsoon season only due to the prevailing water scarcity in other periods (Fig. 6.9). Sometimes, the intensive rainfall causes serious damages to the crops. The average livestock size of a household is around 5–10 and livestock assets are cows, goats, pigs and hens (Fig. 6.10). Villagers had to sell their livestock to generate extra income. On an average, to sell a cow they earned Rs. 4000–10,000/-, from a goat Rs. 2000–3000/-, from a pig Rs. 6000–8000/- and from a hen Rs. 150–200/-. During the field survey, it was found that the people of North Khairbari and Dumchi Rava Basti villages cultivate *kochu* (a root vegetable) under the tree shades. They sell it at Rs. 40/kg in local markets. Another important source of livelihood is betel nut plantation. Betel nut is a familiar profitable cash crop farming in this region, and they can earn Rs. 10,000–12,000/- per 100 trees in a year (on an average). During the field survey, the villagers were asked about the reasons behind the growing popularity of betel nut plantation; one of them informed that:

If I plant trees like Sal, Jarul, Teak etc. which are timber based, I have to take official permission from the forest department to cut the tree for selling or other purposes and often I will not get permission of it. But if I plant betel nuts at once, I can earn money year after year by selling its fruits, without taking permission from the forest department.

Significantly, with the passage of time, the forest villagers are forced to depend on non-forest-based activities. As a result, the originality of the forested landscape has been disturbed by intensive grazing activities and changing the health of forest by planting cash crops.

Fig. 6.10 Different options of livelihood: (**a**) Betel nut farming; (**b**) Vegetable cultivation; (**c**) Livestock rearing of cow; (**d**) Hen; and (**e**) Pig

References

Crona, B., A. Wutich, A. Brewis, and M. Gartin. 2013. Perceptions of climate change: linking local and global perceptions through a cultural knowledge approach. *Climatic Change* 119: 519–531.

Hair, J.F., W.C. Blck, B.J. Babin, R.E. Anderson, and R.L. Tatham. 2006. *Multivariate Data Analysis*. Upper Saddle River, NJ: Pearson Prentice Hall.

Karlsson, B.G. 2013. *Contested Belonging; An Indigenous People's Struggle for Forest and Identity in Sub-Himalayan Bengal*. London: Routledge.

Manandhar, S., V.P. Pandey, and F. Kazama. 2012. Hydro-climatic trends and people's perceptions: case of Kali Gandaki River Basin, Nepal. *Climate Research* 54: 167–179.

Maraseni, T.N., G. Cockfield, and A. Apan. 2005. Community based forest management systems in developing countries and eligibility for clean development mechanism. *Journal of Forest and Livelihood* 4: 31–42.

Prokop, P., and S. Sarkar. 2012. Natural and human impact on land use change of the Sikkimese-Bhutanese Himalayan piedmont, India. *Quaestiones Geographicae Bogucki Wydawnictwo Naukowe, Poznań* 31 (3): 63–75. https://doi.org/10.2478/v10117-012-0010-z.

Ray, S. 2013. *Transformations on the Bengal Frontier: Jalpaiguri 1765-1948*, 19–42. London: Routledge.

Sam, K., and N. Chakma. 2021. *Transformation of Forested Landscape in Bengal Duars: A Geospatial Approach. Spatial Modeling in Forest Resources Management; Rural Livelihood and Sustainable Development, Environmental Science and Engineering*, 553–566. Springer. ISBN 978-3-030-56541-1 ISBN 978-3-030-56542-8 (eBook).

Chapter 7
Adaptation Attitudes of the Forest Villagers

Abstract Adaptation attitudes of the forest villagers implies their responses towards resilience building to cope with the changing climatic conditions in Bengal Duars. This chapter intends to identify the interlinkages between people's concern or belief regarding adaptation and the impact of livelihood assets on their adaptation attitudes. Determination of the influencing factors and the processes responsible towards adaptation are also important issues to give emphasis in this study.

Keywords Adaptation Attitudes · Resilience Building · Forest village

7.1 Conceptual Model of Adaptation

Adaptation is considered as an evolutionary process of adjustment towards any kind of changing situation in the surrounding environment. Therefore, it is important to make strategies to deal with the associated risk and danger. Adaptation is defined as 'response to actual or expected climatic stimuli and their effects or impacts' (Smit and Pilifosova, 2001). The effects and impacts of certain events depends on the way a system is structured and worked. To understand people's adaptation, it is very important to understand the environmental and societal system in which they are living and surviving. In social science research, environmental value and environmental perception always grasp academic consideration (Capstick et al., 2016; Ziegler, 2017).

In Chap. 6 of this book, socio-economic profile of the households living in highly vulnerable areas, their perceptions, struggle in life and livelihood dynamics have been explored in detail. In this Chap. 7, a conceptual model (Fig. 7.1) has been developed through the conceptualisation of the interaction with the impacts of climatic and non-climatic changes, perception of the villagers, and their adaptation strategies on the basis of information collected during the field survey (2015–18). A total number of 21 adaptation choices were identified as personal and institutional adaptation strategies (Fig. 7.2). The actual adaptation was measured through counting of the strategies which are already practiced by the villagers based on the total

© The Author(s), under exclusive license to Springer Nature Switzerland AG 2021
K. Sam, N. Chakma, *Climate Change in the Forest of Bengal Duars*,
SpringerBriefs in Environmental Science, https://doi.org/10.1007/978-3-030-73866-2_7

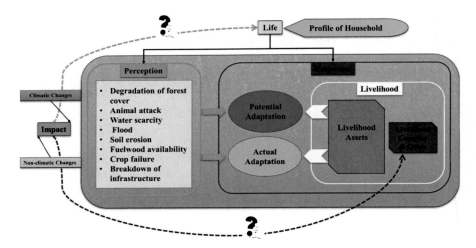

Fig. 7.1 Conceptual Model; Part-I: Impact of changes (climatic and non-climatic) on life and livelihood of people

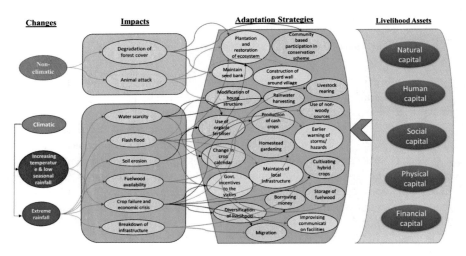

Fig. 7.2 Conceptual Model; Part-II: Describes detail linkages with (**a**) perceived changes (climatic and non-climatic) and adaptation options, (**b**) impact of livelihood assets on adaptation strategies

number of options selected as potential. The 5-point and 3-point Likert scales were preferred to quantify the dependent and independent variables. To test the aforesaid model, the following hypotheses (H) are taken into consideration:

- *H1: The people who are concerned about the climate change and biophysical changes have better potential and actual adaptation*
- *H2: Socio-economic profile of a household has significant influence on adaptation attitudes*
- *H3: Villagers who have better livelihood assets have better adaptation attitude*

The conceptual model's variables and reliability have been stated in Table 7.1. Three separate regression models have been executed by testing the hypothesis to determine whether they are correlated with potential or actual adaptation or not. On the other hand, villager's adaptation capacity has been judged based on livelihood assets to test the abovementioned hypotheses.

7.2   Judgement of Adaptation Attitudes of the Villagers

Adaptation attitudes of the forest villagers are judged on the basis of their potentiality and actual capability. The multi-linear regression method is executed to find out different drivers or factors which determine the actual and potential adaptations of the forest villagers. This analysis will also help to understand the reasons for their responses towards life and livelihoods in the study areas. Judgement of potential and actual adaptations of the villagers and influencing variables are given in Table 7.2.

7.2.1   Potential Adaptation

The potential adaptation of the villagers has been estimated based on the selected options in order to judge their potentiality. Belief and concerns are significantly correlated with the potential adaptation in most cases (H1). It signifies that people with better concern about climate and biophysical changes wish to adapt more strategies. As the old (1st) generation family members were mostly engaged in agriculture and forestry, age is found to be positively associated with the adaptation (H2). The villagers who are dependent on the forest resource have better potential adaptation concern (H2) because of good realization about the consequences happening in the forest that further motivate them to adapt more strategies. Beside these natural, human, and social capital play a significant role in determining people's attitude in this regard (H3). Overall, the model variance explained 65% of the dependent variable for potential adaptation (Table 7.2).

7.2.2   Actual Adaptation

From the analysis, it has been observed that there is no effect of belief or attitude for climate change and biophysical attributes on actual adaptations of the villagers (Table 7.2). Most of the variables are negatively associated, that is, villagers have high apprehension about those facts but the actual implementation or practice of strategies are not visible (H1). Factors like forest dependency and economic status

Table 7.1 Model's variables with the measurement of reliability

Construct	Variable	Scale	Alpha	Eigen value	Factor loading	Mean
Climate change concern	Longer Summer (LS)	Observed/Felt by own = 3; Listen from others = 2; No Idea = 1	1	5.6	0.937	2.67
	Warmer Summer (WS)				0.561	1.89
	Less Cool/Warmer Winter (LCW)				0.937	2.45
	Overlapping of Season (OS)				0.811	2.6
	Extreme Rainfall (ER)				0.849	2.7
	Overall low rainfall (OLR)				0.943	1.09
	Unconventional Formation of Fog (UF)				0.943	1.13
Biophysical concern	Tree Species (TS)	Increase/Improve = 3; Decrease/Degrade = 2; No Idea = 1	0.9	4.82	0.768	1.58
	Faunal Species (FS)				0.943	1.86
	Forest Cover (FC)				0.943	1.79
	Forest Health (FH)				0.805	1.69
	Encroachment (EN)				0.787	1.01
	Grazing (GR)				0.844	1.05
	Illegal Poaching and Filling of Trees (IP_TF)				0.694	2.82
Characteristics of the forest villagers	Age	>65 = 1st generation, 25–65 = 2nd generation, <25 = 3rd generation	0.7	3.57	0.76	2.42
	Gender	Male = 1, Female = 2			0.4	1.39
	Education	Primary = 1, Secondary = 2, Higher Secondary = 3, Graduation = 4			0.42	1.7
	Economic Status	BPL = 1, APL = 2			0.71	1.05
	Dependency on forest resource	3-point scale (High = 3, Low = 1)			0.68	2.57

Construct	Variable		Scale	Alpha	Eigen value	Factor loading	Mean
Livelihood assets	1. Natural capital	Diversity of tree species	3-point scale (High = 3, Low = 1)	0.9	7.5	0.92	2.7
		Revenue from cash crop				0.64	2.4
		Land ownership				0.67	2.1
		Diversity of crop				0.89	2.3
		Proportion of land under cultivation				0.74	2
	2. Human capital	Number of family members	3-point scale (High = 3, Low = 1)			0.88	2.7
		Level of education				0.41	1.8
		Occupational diversity				0.45	2
	3. Social capital	Faith on community	3-point scale (Always = 3, Never = 1)			0.46	2.6
		Supportiveness of community				0.55	2.8
		Active participation in local committee				0.41	1.7
	4. Physical capital	Valuation of assets/equipment of household	3-point scale (High = 3, Low = 1)			0.75	1.92
		Access to fuelwood	3-point scale (Always = 3, Never = 1)			0.83	2
	5. Financial capital	Household budget	3-point scale (High = 3, Low = 1)			0.91	2.1
		Number of income sources				0.47	1.95
		external financial support	3-point scale (Always = 3, Never = 1)			0.4	2

(continued)

Table 7.1 (continued)

Construct	Variable	Scale	Alpha	Eigen value	Factor loading	Mean
Potential adaptation attitudes	Personal	5-point scale (Always = 5, Sometime-3, Never = 1)	0.9	9.25	0.45	2.9
		Modification of house structure			0.9	4.6
		Rainwater harvesting			0.56	1.39
		Use of organic fertilizer			0.87	4.2
		Change in crop calendar			0.78	4.1
		Livestock rearing			0.91	4.5
		Production of cash crops			0.91	4.5
		Homestead gardening			0.87	4.07
		Cultivating hybrid crops			0.89	3.92
		Use of non-woody sources			0.92	3.21
		Storage of fuelwood			0.86	4.21
		Diversified livelihood			0.92	4.57
		Migration			0.9	4.64
		Borrowing money from money lender			0.95	4.31
	Institutional	Plantation and restoration of the ecosystem	0.7	3.9	0.61	3.14
		Maintenance of seed bank of indigenous varieties			0.78	1.84
		Construction of guard wall around villages	5-point scale (Very dissatisfied = 1, Very satisfied = 5)		0.9	3.57
		Early warning of storms/ hazards			0.88	2.41
		Maintenances of local infrastructure			0.93	2.28
		Improvising communication facility			0.94	2.71
		Government incentives to the victims			0.93	2.5
Actual adaptation attitudes		Aggregate of total number of adaptation options practiced by the forest villagers	Count number of variables ranges from 1 to 21	N.A	N.A	N.A	9.13

Table 7.2 Factors affecting potential and actual adaption attitudes of the forest villagers

Domain	Variable	Potential adaptation	Actual adaptation
Belief/attitude	Climate change concern	0.265 (0.082)**	−0.273 (0.081)
	Biophysical concern	0.24 (0.056)**	−0.114 (0.055)
Norms	Environmental policy	0.206 (0.071)**	−0.202 (0.070)
Characteristics of forest villagers	Age	0.624 (0.235)*	0.356 (0.271)
	Gender	0.016 (0.011)	0.024(0.030)
	Education	0.137 (0.026)	0.019(0.124)
	Economic status	0.168 (0.047)	0.037 (0.029)**
	Dependency on forest resource	0.059 (0.010)*	−0.113 (0.024)*
Livelihood assets	Natural capital	0.134 (0.061)*	0.121 (0.04)**
	Human capital	0.152 (0.013)*	0.112(0.013)
	Social capital	0.319 (0.026)**	0.284(0.016)
	Physical capital	0.53 (0.034)	0.204 (0.061)
	Financial capital	0.671 (0.045)	0.451 (0.026)*
R^2		0.651	0.537
Adjusted R^2		0.61	0.501

Source: Field Survey
**Significant at 0.01 level, *Significant at 0.05 level
Rows reflect independent variable with beta (β) and standard error shown within parenthesis.
Columns represent dependent variable. Shaded boxes indicate factor significant at different level.

are significantly associated with actual adaptation. Dependency on forest is negatively associated with actual adaptation (H2). The villagers who have more dependency on forest resources have limited options of livelihood, and therefore, as a result their economic condition are not so good.

On the other hand, the villagers who do not have not so much dependency on forest-based resources and have more dependency on agriculture-related activities, cash crop plantation and livestock rearing are found with a better economic condition. It has been found that the people with better economic status and less dependency of forest have improved their actual adaptation attitude (H2) by practicing more strategies. Earning money from the natural capital (cash crop betel nut) improves the financial capital of the households. Beside these, wage labour and migrated labour also contribute to improve the actual adaptation of forest villagers. Therefore, the natural and financial capital have a significant role in actual adaptation (H3). Altogether, the model explained 53% of the variance of the dependent variables in case of actual adaptation (Table 7.2).

7.3   Discussion

From the analysis, it has been found that the R^2 value of potential adaptation is high in comparison to the actual adaptation. Forest villagers in Bengal Duars are very well concerned about the changes in their surroundings but the situation and circumstances of their living make survival difficult. This is the reason that most of them are still struggling in the studied areas for their life and livelihoods. The recent trend of out-migration signifies the crisis and lack of involvement in forest-based activities. If it ensues for a long period of time, then villages may be depopulated in future. Therefore, it is very important to restore the landscape by considering both ecological as well as economic benefits of the inhabitants.

References

Capstick, S.B., N.F. Pidgeon, A.J. Corner, E.M. Spence, and P.N. Pearson. 2016. Public understanding in Great Britain of ocean acidification. *Nature Climate Change* 6 (8): 763–767.
Smit, B., and O. Pilifosova. 2001. Adaptation to climate change in the context of sustainable development and equity. ed. IPCC. *Impacts, Adaptation and Vulnerability Contribution of Working Group II Third Assessment Report of the Intergovernmental Panel on Climate Change.* Cambridge: Cambridge University Press.
Ziegler, A. 2017. Political orientation, environmental values, and climate change beliefs and attitudes: An empirical cross country analysis. *Energy Economics* 63: 144–153.

Chapter 8
Recommendations and Conclusion

Abstract In developing countries like India, effective planning and strategies for vulnerable population are very important to ensure the safety of their life and livelihoods to combat with the climatic risk and at the same time to save rich forest resources in the region. This chapter explores existing challenges, summarizes the findings and proposes recommendations for furtherance of the forested landscape.

Keywords Planning · Strategies · Challenges · Recommendations

8.1 Introduction to REDD+

The United Nations Framework Convention on Climate Change (UNFCCC) introduced the agenda of 'Reducing emission from deforestation and forest degradation' (REDD). Later on, the revised version of REDD+ was adopted with reference to the specific four key elements 'for forest conservation, sustainable management of forests and enhancement of forest carbon stocks in developing countries'. The key elements are as follows:

1. National Forest Monitoring System (NFMS)
2. Safeguard Information System (SIS)
3. Forest Reference (emission) Level (FRL)
4. National Strategies /Action Plan (NS/AC)

The Paris Agreement on Climate Change (2015) in Article 15 has mentioned the role and contribution of REDD+. In the 11th session of Conference of Parties (COP) to the UNFCCC, the final decision was adopted at COP16 at Cancun in December 2010 by encouraging developing countries to undertake REDD+ activities. The three phases mechanism of REDD+ ensures to provide financial incentives by supporting 'result-based actions'. It has been decided to implement REDD+ strategy in India too.

The National Forest Monitoring System (NFMS) in India has been monitoring forest resources since 1987 using Landsat MSS data. After 2001, through the advancement of space technology, LISS-III data with spatial resolution 23.5 m has been used for forest monitoring purposes. The Forest Survey of India has published the State Forest Report with a year's interval on a regular basis. However, there are inconsistencies in the reports compared to the ground truth result and high-resolution image analysis regarding this. There is lack of a strong permanent monitoring system for carbon stock assessment in India. Recently, emphasis has been made on it, and a separate chapter of carbon stock was incorporated in India's State Forest Report, 2019.

The Cancun Agreement (1/CP.16) includes various principles concerning safeguards in information system. Till now, India does not have any efficient system for providing information related to environmental and social safeguards. A well-defined system with a set of indicators such as forest governance, rights of indigenous people and effective involvement of stakeholders are necessary for monitoring purposes (SFR 2009). Gradually from 2011, the Forest Survey of India (FSI) has published about forest villagers and their dependency on forest in detail.

For developing nations, it is most challenging to set the Forest Reference Level (FRL) with historic and present information about forest area and carbon emission. Spatial and non-spatial data of drivers are required for FRL estimation. Low-resolution free remote sensing data like Landsat are used by most of the countries but are not very effective for assessment of forest degradation (Poudel and Poudel 2018) However, there are significant challenges to avail consistent historical information of carbon stock (Ravindranath et al. 2012). A report submitted to the UNFCCC, it has been mentioned that 49 million tonnes of CO_2 equivalent as a FRL for the year 2000–2008 (MoEFCC 2018).

National Action Plan on Climate Change (NAPCC) was released by then Prime Minister Dr. Manmohan Singh in 2008. It aimed to reduce vulnerability to climate change through multilevel integrated strategies in the context of climate change. The NAPCC outlines eight national missions under which the National Green Mission Plan addressed on the indispensable role of forest to maintain ecological balance and biodiversity. Under the 'Green India' scheme, it is targeted to increase 5 million hectares of forest/tree cover and improve their quality. Other targets are to generate forest-based livelihood income for 3 million households living in and surrounding the forest and also to enhance annual sequestration of CO_2 by 50–60 million tonnes in 2020. The mission of the National Clean Energy Fund (NCEF) is to harness the use of renewable energy sources in order to reduce dependency on fossil fuels for combating global warming and climate change.

India has robust strategies for the management of forest resources. India has practiced dissimilar forest management and conservation policy in Colonial and post-Colonial periods. During the Colonial period, British administration exploited forest resources in the name of commercialisation without conservation. Most of the policies were given importance on forest resources rather than forest-dependent people (Balaji 2002). In the post-Colonial period, the forest policy of 1988 recognised the role of local people for protection and management of forest to improve the community livelihood. This policy was further revised in 1990 to a community-based management system named as Joint Forest Management (JFM). JFM is based on the principle of 'Care and Share'. Presently, India has more than 118,213 JFM committees including around 20 million people to manage more than 25 mha of forest cover (MoEFCC 2018). The recently published State Forest Report in 2019 has incorporated many technical details, but the role and effectiveness of JFM as a powerful national-level strategy has not been evaluated properly.

8.2 Major Challenges of Bengal Duars

As per the Constitution of India, 'forest' appears in the concurrent list; therefore, both the state and the central governments have authorities on forest resources. The federal government formulates some policies and programmes, and the state government plays a significant role to implement and manage forestry. On the other hand, climate change has become a harsh reality globally but its intensity and impacts are viable at local level also. India has already taken initiatives to implement REDD+ strategies by following the guidelines of UNFCCC, but still several challenges are present at the ground level. These challenges are discussed here in the context of Bengal Duars.

8.2.1 Challenges to National Forest Monitoring System

The Forest Survey of India, Government of India publishes annual reports at a regular basis stating the statistics regarding area under forest and non-forest covers of the districts. On the other hand, each forest division in several states are also preparing their own reports, although in most of the cases in spite of giving micro-level details about spatial gain and loss of forest covers, only an average value is mentioned. This is the case of several published reports of forest divisions in Bengal Duars also. Further, much importance has been given on economic cost-benefit analysis through the collection of revenues, whereas rate of deforestation and reforestation are not mentioned for the succeeding years. However, use of high-resolution satellite images, ground-based assessments and validation are very important to investigate the reality and changes occurring in forests. During preparation of planning at micro level plot-wise land cover maps with the participation of the local community are very important and effective to manage the forest resources, which was also absent in Bengal Duars.

8.2.2 Challenges to Safeguard Information System

This region is a hub of indigenous communities; their knowledge is vital for conservation of nature and natural systems as well as to enhance societal and environmental benefits. During field survey, it was observed that collection of forest products by

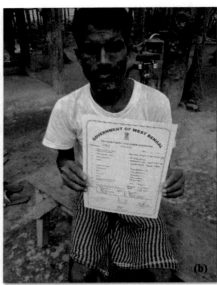

Fig. 8.1 (**a**) A villager in Dumchi Rabha Basti shows 'patta' (Forest right ownership certificate); (**b**) A villager in Kodal Basti shows forest occupational certificate

forest dwellers other than wood (non-timber products like fruits, leaves, etc.) was restricted by the forest departments. Till now villagers are struggling for their land rights, even some of them do not have a legal authorized land rights certificate called 'patta' (Fig. 8.1). Forest villagers are suppressed and facing hurdles to practice and utilize their traditional knowledge for economic and environmental benefit.

8.2.3   Challenges to Monitor Forest Reference (Emission) Level

Regular monitoring of Forest Reference Level has played a significant role to know the current status of the landscape. In Bengal Duars, there is a lack of information about Forest Reference Level due to the unavailability of micro-level forest information system. As this region has experienced severe deforestation, use of modern technology to monitor the landscape, therefore, is very essential.

8.2.4 Challenges to Implement National Strategies or Action Plan

As per provision of the National Forest Policy (1988), the Government of India has formulated a framework for protection, regeneration and development of forest cover through the active involvement of local communities named as Joint Forest Management (JFM) programme. Under this scheme, there are Forest Protection Committees or Eco Development Committees to make cooperation between stakeholders and government. However, these committees are not straightforward outcome, but a result of politician, bureaucrats and non-governmental organization. In 1996, this programme was extended to the protected areas like national parks and wildlife sanctuaries. In North Bengal, JFM is still evolving due to the presence of tea gardens close to the forest. Presence of large protected areas accompanied with forest villagers make things more complicated (Gupta 2005).

The studied region always came in front page headlines regarding human encroachment, illegal logging and poaching activities (Prokop and Sarkar 2012; Das 2012; Bhattacharyya and Padhy 2013; The Telegraph 2006, 2008; The Telegraph 2018). The negative attitudes of forest managers to forest dwellers by ignoring their legitimate rights over their ancestral land in Buxa Tiger Reserve (BTR) (Das 2012) is present till date. The Wildlife Act 1972, Section 29 allows such activities in forest which are not destructive for wildlife. Elsewhere, Section 24 (2) (c) permits continuation of forest right within sanctuary at the discretion of the collector. The act of allowing participation and benefit of sharing has not been fully utilized by forest villagers, because the authority assumed that the activities of the forest dwellers in protected areas is necessarily destructive in nature. Therefore, the virtue of Section 29 has a role to play and resulted in a ban to collect non-timber forest products from protected areas by forest authority. During the field observation, discrepancies were observed in 28th Mile Basti of Buxa Tiger Reserve in 2015; during previous visits, villagers told that they used to collect some non-timber products from forest and also sell it to the buyers (Fig. 8.2). But when this forest *basti* was revisited again in 2018 villagers confessed to us that due to some illegal activities the forest department now totally banned collecting those non-timber products. Even in some protected areas of Buxa Tiger Reserve, human settlement extended up to the core areas and its fringes for a long time, and the authorities tried to rehabilitate those villagers but they were not convinced without proper assurance and facilities.

8.3 Suggestive Measures

The forest cover of Bengal Duars is still degrading despite having vigorous National Forest Policy. In such a situation, it is very important to restore this landscape by following some ecosystem-based adaptation strategies. In restoration strategy, ecosystem-based approach is dynamic for long-term future adjustment with the

Fig. 8.2 Collection of non-timber products by forest villagers in 28th mile Basti (2015), and local names are mentioned in the individual picture

Fig. 8.3 Identification and planning of restoration opportunities in different sites of Bengal Duars

ecosystem (IUCN and WRI 2014). This approach follows some organized principles and these are as follows:

1. *Focus on Landscape:* At first, it is important to classify individual sites as per the condition such as protected area, primary forest, secondary forest, agroforestry, well-managed plantation, agricultural land, and non-forest area. (Fig. 8.3).

2. *Restore Functionality:* Hazards like floods, landslides, and soil erosion are common in this region as it is located in the piedmont zone. In order to restore the landscape, watershed wise rate of erosion must be checked at a regular basis. Then, further based on the scale of vulnerability and priority of functionality, the landscape needs to be restored by using the principle of 'back to the original' vegetation and other strategies. Surveyed villagers have mentioned some references of native tree species like Sal *(Shorea robusta)*, Chilauni *(Schima wallichii)*, Tatri *(Dillenia pentagyna)*, Lali *(Phoebe species)* and Tun *(Cedrela*

toona). They also proposed to construct more ponds, depressions and lakes for both environmental as well as economic benefits.

3. ***Allow for Multiple Benefits:*** Forest is a valuable resource that provides multiple benefits to the environment as well as organisms and makes the ecosystem healthy. As this region is erosion-prone in nature, close canopy coverage in the upper catchment of the river needs to be protected for down-streaming water supply, reducing erosion and sequestering large amounts of carbon. Plantation can also be practiced around the agricultural field, and there should be limitations to plant cash crops per household, because cash crops like betel nuts are changing the hydrological character of soil and accelerates run off to increase the rate of erosion by reducing infiltration rate in forest sites (Cheng et al. 2008). Native trees with more canopy cover have potentiality to reduce overflow (Fig. 8.4). It is also important to encourage the growth of more medicinal plants and diversified utilization of common property resources to generate more non-farm-based activity. Moreover, enhancement of eco-tourism and community participation, the linking between cultural and natural sites by using modern technology like mobile apps can also be an alternative benefit.

4. ***Leverage Suit of Strategies:*** The use of modern technological assistance for real-time monitoring of the activities going on in the forest will help further to stop illegal felling and poaching of animals. The high-resolution sensor tool is also beneficial to detect the health of the vegetation.

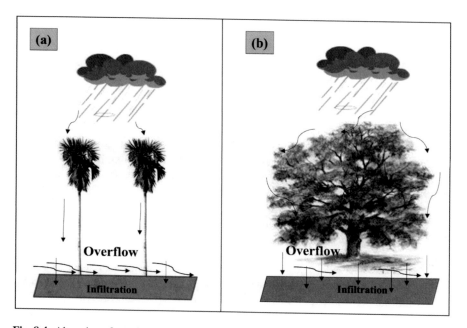

Fig. 8.4 Alteration of overflow and infiltration character in case of (**a**) betel nut plantation and (**b**) native trees

5. *Involvement of Stakeholder:* The active participation of local group and community like Forest Protection Committees (FPC) is vital for discussion, formulation and implementation of strategies. The members of those committees in forest villages can be chosen by different generations of the households as different generations have different experiences and knowledge to share in the scope of development for their own area. Moreover, the restoration process should respect their right on forest, land and resources without destruction.

6. *Tailor to Local Condition:* Each and every landscape is unique in context, and there is no such 'one size fits all' model in the landscape restoration process. Thus, evaluation and assessment of the societal, economic and ecological context of a particular region is very essential.

7. *Avoid Further Deduction of Forest Cover:* A regular monitoring system is needed for the stability of forest cover. Community-based labour force can be appointed for plantation protection activities, side-by-side use of modern sensor-based technology can also assist accordingly. Villagers must monitor and assess the activities of the forest protection committees and can also report to the forest department on a regular basis. However, to avoid nuisance of illegal activity by villagers, only female members of the households may be allowed to collect fuelwood, fodders and non-timber forest products, as they have relatively less muscle power than male.

8. *Adaptively Manage:* With the change of environmental condition, human knowledge, technology and social values, local people should make an adjustment to the progress of restoration strategies.

8.4 Conclusion

Forest is a natural air cooler in not only providing internal comfort to the people dependent on it for their life and livelihoods but also extending external urge to the natural environment for comfortable living. The forest is now one of the biggest vulnerable resources all around the world mostly due to anthropogenic duress as the recent incidents of forest fire happened in the Amazon, Australia, India, etc. Hence, the importance of forest becomes prioritized when incidences like unpleasant weather due to climate change and struggle with air and water pollution make our life become suffocative.

The people in the forest of Bengal Duars are struggling a lot for survival. Through the study, the following conclusions are emerged:

First, in the Bengal Duars, climate change is now a serious issue in the degraded forested landscape. As a result, associated vulnerability has increased as related with extreme flood events, overlapping of season, attack of animals, distress in life and livelihood provoking people for out-migration.

Second, the socio-economic conditions of the households are very weak. Most of them are struggling for their forest rights, some are working as small and marginal farmers. People are basically more dependent on non-forest-based activities though they are living within the forest and are engaged in cash crop (betel nut) plantation as an alternative livelihood.

Third, they have very good perception about climatic and non-climatic changes of their surroundings, but their actual adaptive capability is not up to the mark due to socio-economic and infrastructural hindrances. This does not seem to be good for future threats in connection with the incidence of climate change.

Fourth, the central and state governments have formulated several developmental programmes but villagers are mostly dissatisfied about proper implementation and outcomes of those schemes.

Finally, an integrated research and development approach needs to be adapted by the villagers through the collaboration of bio-engineering and social engineering by government role and support regarding this. The knowledge of indigenous communities may help in understanding effective treatment of degraded land with the fusion of modern technology. This may result in dismissing the vicious cycle of land degradation, non-forest-based dependency on livelihood and out-migration of forest villagers. For the policy makers before formulating any programme and policy exercise of ROAM (Restoration Opportunities Assessment Methodology) model can be useful to integrate local knowledge and best available science as a cost-benefit carbon restoration framework proposed by the IUCN and WRI. ROAM has been used by many countries such as Rwanda, Mexico, Uganda, etc. After realizing the urgent needs for restoration in India, ROAM is using FLR (Forest Landscape Restoration) areas through a sub-national assessment in Uttarakhand.

The ROAM model contains three phases of assessment; these are as follows:

1. ***Preparation and Planning:*** This phase is likely to involve several discussions with direct and indirect stakeholders, and other groups to understand the problems and challenges. Then, it is needed to define national and sub-national level objectives and goals for the better assessment.
2. ***Data Collection and Analysis:*** This phase deals with the type of data required for the analysis. Most of the data should be spatial in nature to be mapped easily. Other types of data particularly related to programmes, policies, socio-economic and financial conditions are also useful for analysis.
3. ***Result Validation and Recommendation:*** After presentation of findings of a proposal, feedback is required from the senior departmental staff, national-level experts and local stakeholders. Finally, their critical assessment is needed to be brought together to make valid recommendations for a particular landscape.

References

Balaji, S. 2002. *Forest Policy in India – in Retrospect and Prospect, IUFRO Science/Policy Interface Task Force regional meeting.* Chennai: M.S. Swaminathan Research Foundation.
Bhattacharyya, M.K., and P.K. Padhy. 2013. Forest and wildlife scenarios of Northern West Bengal, India: a review. *International Research Journal of Biological Sciences* 2 (7): 70–79.

Cheng, J.D., J.P. Lin, S.Y. Lu, L.S. Huange, and H.L. Wu. 2008. Hydrological characteristics of betel nut plantations on slopelands in central Taiwan / Caractéristiques hydrologiques de plantations de noix de bétel sur des versants du centre Taïwan. *Hydrological Sciences Journal* 53 (6): 1208–1220. https://doi.org/10.1623/hysj.53.6.1208.

Das, B. K. 2012. *Losing Biodiversity, Impoverishing Forest Villagers: Analysing forest policies in the context of flood disaster in a national park of Sub Himalaya Bengal, India*. Occasional Paper 35, Kolkata: Institute of Development Studies.

Gupta, K. 2005. *The Political Economy of Forest Management; Importance of Institution and Social Capital*. New Delhi: Allied Publishers Pvt.

IUCN and WRI. 2014. *A guide to the Restoration Opportunities Assessment Methodology (ROAM): Assessing forest landscape restoration opportunities at the national or sub-national level*. Working Paper (Road-test edition). Gland, Switzerland: IUCN.

MoEFCC. 2018. *Natonal REDD+ Strategy India*. Ministry of Environment, Forest and Climate Change, Government of India.

Poudel, M., and A.K. Poudel. 2018. Forest reference levels in the Hindukush Himalaya: a comparative overview. *Journal of Forest and Livelihood* 17 (1): 90–110.

Prokop, P. and S. Sarkar. 2012. Natural and human impact on land use change of the Sikkimese-Bhutanese Himalayan piedmont, India. *Quaestiones Geographicae* 31(3): Bogucki Wydawnictwo Naukowe, Poznań. 63–75. https://doi.org/10.2478/v10117-012-0010-z. ISSN 0137-477X.

Ravindranath, N.H., N. Srivastava, I.K. Murthy, S. Malaviya, M. Munsi, and N. Sharma. 2012. Deforestation and forest degradation in India – implications for REDD+. *Current Science* 102 (8): 1117–1125.

SFR. 2009. *The State of Forest Report*. FSI, MoEF, Government of India.

The Telegraph. 2006. *Two courts to save forest*. 15.02.2006. West Bengal.

———. 2008. *Timber Smuggling on train thrives- Forest minister alleges GRP mafia nexus*. 06.08.2008. West Bengal.

———. 2018. *More cameras for forest*. 16.02.2018. West Bengal.

Glossary

Basti Village hamlet located in forest areas.

Bigha A measure of land about 1/3 of an acre.

Duars/Dooars/Doars Doors or corridors to access Bhutan from alluvial plains, located in between river Tista and river Dhansiri.

Jhum Shifting cultivation or slash and burn cultivation.

Jungle A land covered by dense forest and tangled vegetation.

Mouza A local administrative unit corresponding to a specific area of land with or without settlement.

Patta An official recording of land holds by a family or household.

Tarai/Terai A region of emergence of streams composed with finer sediment located at the foot of the Siwalik Himalaya.

Taungya A traditional method of tree cultivation practised by the hill people in Burma (now Myanmar) where crops are cultivated between the tree lines.

© The Author(s), under exclusive license to Springer Nature Switzerland AG 2021   71
K. Sam, N. Chakma, *Climate Change in the Forest of Bengal Duars*,
SpringerBriefs in Environmental Science, https://doi.org/10.1007/978-3-030-73866-2

Printed in the United States
by Baker & Taylor Publisher Services